CW00339438

Post-Suburban Europe

Also by Nicholas A. Phelps

MULTINATIONALS AND EUROPEAN INTEGRATION: Trade, Investment and Regional Development

FOREIGN DIRECT INVESTMENT AND THE GLOBAL ECONOMY: Corporate and Institutional Dynamics of Global-localisation (*editor*)

THE NEW COMPETITION FOR INWARD INVESTMENT: Companies, Institutions and Territorial Development (*editor*)

Also by Nick Parsons

EMPLOYEE PARTICIPATION IN EUROPE
REINVENTING FRANCE (*editor*)

Post-Suburban Europe

Planning and Politics at the Margins of Europe's Capital Cities

Nicholas A. Phelps
University of Southampton, UK

Nick Parsons
Cardiff School of European Studies, Cardiff University, UK

Dimitris Ballas
University of Sheffield, UK

Andrew Dowling
Cardiff School of European Studies, Cardiff University, UK

First published 2006 by
PALGRAVE MACMILLAN
Houndmills, Basingstoke, Hampshire RG21 6XS and
175 Fifth Avenue, New York, N.Y. 10010
Companies and representatives throughout the world

PALGRAVE MACMILLAN is the global academic imprint of the Palgrave Macmillan division of St. Martin's Press, LLC and of Palgrave Macmillan Ltd. Macmillan® is a registered trademark in the United States, United Kingdom and other countries. Palgrave is a registered trademark in the European Union and other countries.

ISBN-13: 978–0–230–00212–8 hardback
ISBN-10: 0–230–00212–9 hardback

This book is printed on paper suitable for recycling and made from fully managed and sustained forest sources.

A catalogue record for this book is available from the British Library.

Library of Congress Cataloging-in-Publication Data
Post-suburban Europe : planning and politics at the margins of Europe's
   capital cities / Nicholas A. Phelps . . . [et al.].
      p.   cm.
Includes bibliographical references and index.
ISBN 0–230–00212–9 (cloth)
1. Suburbs—Europe. 2. City planning—Europe.   I. Phelps,
N. A. (Nicholas A.)
HT352.E85P67 2006
307.76094—dc22   2006045225

10   9   8   7   6   5   4   3   2   1
15   14   13   12   11   10   09   08   07   06

Printed and bound in Great Britain by
Antony Rowe Ltd, Chippenham and Eastbourne

Contents

List of Figures

List of Table

Preface

The research contained in this book has its origins in an interest sparked over a decade ago. In the summer of 1993 Nicholas Phelps was employed on a series of economic development projects commissioned by Croydon Council. At that time, major institutions in Croydon – such as the Council – were experiencing some introspection about the future prospects of the local economy. Croydon was, and to some extent still is, the brunt of numerous jokes regarding its banality in the popular media. Yet even casual inspection suggested this *was* an interesting and curious place. Since Peter Saunders' classic study of *Urban Politics: A Sociological Interpretation*, Croydon, along with many other seemingly banal places at the edge of major cities, had been overlooked in terms of what it might tell us about contemporary urbanisation and economic development. Not long after 1993, and as part of a series of initiatives aimed at transforming the physical and institutional fabric of the borough, Croydon Council was prominent in establishing a unique European network of edge cities. It was the very idea of such a network that first sparked an interest in researching the politics, planning and economy of post-suburbia in a specifically European setting.

However, the garnering of serious interest from other academics in the study of seemingly banal post-suburban places in a European setting has been a difficult process. On several occasions the mere mention of the term 'edge cities' in connection with Europe has evoked much misunderstanding and hostility from fellow academics. Perhaps as a result, it also proved difficult to find a publisher willing to entertain the subject matter. Palgrave publishers are therefore to be thanked for their interest in this interdisciplinary monograph and also for the overall professionalism of their response. Let us make it clear then, although the study of European post-suburbia presented here took for convenience-sake its starting point a recently created European network of edge cities, we are certainly not saying that these can be regarded as European counterparts to those defined in the United States setting by Joel Garreau in his book *Edge City*. As this study makes clear, the use of this term in a European setting was opportunistic and the search for United States-style edge cities in Europe would be futile. Instead, the network of municipalities represented a convenient entry point into exploring the diversity of Europe's emerging post-suburban landscape.

As one might expect of a group of rather invisible or anonymous places – places off the international academic map within urban studies – the research contained in this book derives from only very modest research grant funding. The research did not benefit from the employ-ment of full-time research assistants but has from the outset been a part-time endeavour spread across a network of friends with rather different primary research interests – French industrial relations, micro-simulation, Catalan history, the economic geography of multinational companies. This book no doubt therefore bears the marks of omissions borne from incomplete immersion in the relevant literature in general and the specifics of urban studies literature in each of the national contexts. These omissions may be all the more severe given our inter-disciplinary approach to the task at hand. The empirical substance of the book also bears the marks of the specific remit of the two projects funded by the UK Economic and Social Research Council and British Academy. These two research projects were aimed at understanding the governance of, and role of business interests in, the set of post-suburban municipalities considered in this book. As such, the research reported in this book does have an emphasis on government and quasi-government bodies and businesses in shaping post-suburban Europe as opposed to, for example, land owners, building companies and residents and environmental groups.

Nevertheless, we hope the empirical material presented here provides a relatively complete story of each municipality and, when viewed as a whole, the diversity of post-suburban Europe. We hope that this book can form something of a bridge between disciplinary perspectives and reorient urban studies away from global cities *per se* to a more balanced appreciation of global city-regions and the immense scale of urbanisa-tion in which the centres of global cities are enmeshed.

Nicholas A. Phelps,
Southampton

Nick Parsons,
Cardiff

Dimitris Ballas,
Sheffield

Andrew Dowling,
Cardiff

Acknowledgements

A book based upon research spanning five localities across Europe has necessarily incurred many debts – personal and professional – along the way. Our first thanks must be to the Economic and Social Research Council and the British Academy for funding the research on which this book is based. We would also like to thank all those who kindly gave their time to be interviewed and invariably talked with insight and verve regarding their respective localities.

Teresa Connelly and Graham Sadler's house in Oxted was a convenient and immensely hospitable retreat from research in Croydon. John Whittington and colleagues at Croydon Council deserve special thanks for their continued interest in and help with this research. Thanks are also due to Donald McNeill for his interest and involvement at the outset of this research odyssey.

In Finland, thanks are due to Mervi Ilmonen at YTK at Helsinki University of Technology, to Irma Karjaleinen and Matti Kokkinen at Espoo City Council and to Douglas Gordon at Helsinki City Council for their invaluable assistance and encouragement with this research. Thanks also to Dave Milne at University of Leeds and to two years worth of second-year Geography and to the Geography and Transport Studies students for contributing to two enjoyable field trips to Helsinki and Espoo in 2003 and 2004.

Thanks to Sterios Konstantinou, Yiannis Yiatakis and the extended Ballas family – Giorgos, Tania, Nikos, Chloe, Vassili and Roula – for their generous hospitality and introduction to Polykatokia living and the Athens roadscape.

Thanks to staff at GISA in Getafe and in particular Santiago Fernández Gonzalez for facilitating interviews in Getafe, and to Max de Vries for providing accommodation and excellent company while in Madrid.

Our thanks too to Hervet Mahot at Comex 93 in Seine-Saint-Denis for giving generously of his time and facilitating additional interview sources.

Lyn Ertl at the Cartographic Unit, School of Geography, University of Southampton should be thanked for her expert and speedy production of the diagrams contained in this book. Finally, thanks to Jennifer Nelson formerly at Palgrave publishers for her enthusiastic championing of our initial book proposal and to Gemma d'Arcy Hughes at Palgrave and Vidya Vijayan at Integra for their efficiency in editing and in helping to hone the final appearance of this book.

1
Introduction

> Over the centuries that followed, catastrophic incursions of the
> sea into the land of this kind happened time and again... Little
> by little the people of Dunwich accepted the inevitability
> of their hopeless struggle, turned their backs on the sea,
> and...built to the westward in a protracted flight that went
> on for generations; the slowly dying town thus followed – by
> reflex, one might say – one of the fundamental patterns of
> human behaviour. A strikingly large number of our settlements
> are oriented to the west and, where circumstances permit, relo-
> cate in a westward direction. The east stands for lost causes.
>
> W.G. Sebald, *The Rings of Saturn*

1.1 Californication?

'Post-suburbia' is a term that encapsulates a variety of contemporary
urban forms – 'edge cities' (Garreau, 1991), 'edgeless cities' (Lang, 2003),
'exurbia' (Soja, 2000), 'technoburbs' (Fishman, 1987) – perceived to
be part of the expanding urban fabric of metropolitan regions. Either
at their creation or through the subsequent accretion of commercial
and economic activities to residential development, these spaces are
distinctly *post*-suburban. While post-suburbia remains invested with
much of the ideological elements associated with suburbia its economic
function entails a set of political-economic tensions (Teaford, 1997). It
is some of these political-economic tensions that we seek to explore
in this book when considering the development of five post-suburban
municipalities at the edge of European capital cities.

In tracing the lineage of the postmetropolis, Ed Soja pays brief homage
to one of the oldest known cities – Çatal Hüyük (Soja, 2000). His

analytical excavations of this ancient middle eastern city are a brief prelude, however, to uncovering the most modern of processes and forms of urbanisation exemplified by Los Angeles. Sjoberg cast his net wide to capture the essential traits of the pre-industrial city (Sjoberg, 1960). Manchester and Chicago were the exemplars of the modern industrial city for Engels and Burgess, Hoyt and Park. For Soja and others (Dear, 2003), it is Los Angeles that has come to embody the contemporary highpoint of post-modern urbanisation processes.

Apparently as late as the mid-1990s it was 'still an open question... whether to view Los Angeles as an exceptional case, a persistently peculiar and irreproducible type of city, or as an exemplary, if not paradigmatic illustration of the essential and generalizable features of late twentieth century urbanisation' (Soja and Scott, 1996). Indeed, right from the outset, those planning Los Angeles consciously promoted the metropolis as the norm for future urban development. The Director of the Los Angeles City Planning Commission in 1925, for example, argued that 'the Eastern type of metropolis is unnatural – even wasteful' and that because of 'certain new factors, the Los Angeles (or Western) type of development is normal, economically sound, socially correct and consequently unusually stable' (quoted in Hise, 1997: 43–44).

Sprawling Los Angeles with its multiple centres now represents a byword for the post-war low-density suburban development that can be found surrounding and in between major cities across the United States (US) and North America. It exemplifies the manner in which the city has increasingly been turned inside out (Soja, 2000). Yet the multiple centres of Los Angeles are more balanced than has commonly been recognised. In many instances, from the outset, they embodied something more than simply suburban residential development, including significant manufacturing and service sector economic activities (Hise, 1997). Otherwise, such commercial and economic functions became accretions to initial suburban residential developments. As such, the sprawling edge cities of greater Los Angeles provide some of the archetypal examples of post-suburbia (Garreau, 1991; Teaford, 1997).

In the welter of academic research and popular commentary on Los Angeles, this particular urban sprawl has come to represent both something different and something that is being generalised. Nowhere is this more apparent than in the deliberations of a self-proclaimed 'Los Angeles school' (Dear, 2003). Thus, on the one hand, 'Los Angeles is the first consequential American city to separate itself decisively from European models and to reveal the impulse to privatisation embedded in the origins of the American revolution' (Weinstein, 1996: 22). A few years later, and on the other hand, Soja felt able to assert that 'Los Angeles is a . . . representative window through which one can observe in all their uniquely expressed generality the new urbanisation processes that have been reshaping cities and urban life everywhere in the world over the past thirty years' and that 'what has been happening in Los Angeles can also be seen taking place in Peoria, Scunthorpe, Belo Horizonte, and Kaotsiung, with varying intensities to be sure and *never* in exactly the same way' (Soja, 2000: xvii).

According to this latter view, there is a distinct sense then in which Los Angeles, at a most westerly point in the western hemisphere, has come to represent the final word on urban process and form. Yet, as Sjoberg (1960) noted of the dominant Chicago school in his deliberations on the pre-industrial city, 'many of the recent generalisations of sociologists derived solely from evidence in American society, and then only for a short time-span, are certain to prove inadequate' (Sjoberg, 1960: 2). The impulse to such unifying analyses is strong. Despite the identification of extended metropolitan forms specifically with processes of urbanisation in East Asia (McGee, 1991), they have also been likened to the sort of the US sprawl represented by Los Angeles (Dick and Rimmer, 1998). Thus, 'notwithstanding the very different settlement patterns on which it is being imposed, the new arrangement matches very closely with what Garreau described in *Edge City*' (Dick and Rimmer, 1998: 2304). It is noticeable, however, that such unifying analyses stop short of including the diversity of European experiences of urbanisation. So, for example, the search for US-style edge cities within Europe may itself be rather futile. The term 'edge city' is by now so firmly invested with connotations of the *form* of North American urbanisation as to potentially obscure any valid points of comparison in the *function* of post-suburban developments in different national and continental settings. So, the five case studies of post-suburban development presented in this book, although in appearance rather unlike a US edge city, correspond to at least some of Garreau's (1991) criteria. In this study, we have tried to remain open to points of comparison and contrast in the function and form of post-suburban developments in different settings, although our sympathies lie in contributing to a geographical analysis of post-suburban difference (Fincher *et al.*, 2002).

If Los Angeles represents the high point of contemporary urbanisation, where might one to look for indications of future patterns of urbanisation? One obvious answer is to look straight out westward from Venice beach to . . . the east. Certainly, what has been dubbed the 'Pacific century' appears to be ushering in a rural–urban transition of a scale like none preceding; this in turn driving the construction of urban mega projects. The key point here is that we will have come full circle figuratively and quite possibly analytically. Although it is tempting, as the extract from *The Rings of Saturn* suggests, to observe a unidirectional movement westward, the path of 'progress' might equally be eastward.

In this book we wish to privilege neither west nor east as the appropriate vantage point for observing contemporary patterns of urbanisation and the rise of post-suburbia in particular. Instead the cradle of civilisation – Çatal Hüyük and the ancient cities of Mesopotamia – in the middle ground might offer a metaphor for approaching contemporary patterns of urbanisation. The patterns and processes of urbanisation at the edge of European capital cities are the subject of this book but this is not an argument for returning urban studies to some older European or Middle Eastern point of reference. The idea of the middle ground serves as an analytical metaphor in several respects – the least of these being the problematic contrasts to be drawn between a European middle ground and that of the Far East and North America and the equally problematic notion of westward 'progress'.

Contrasts between east and west can of course be drawn at other lesser geographical scales. Within a relatively circumscribed location in global terms, Mesopotamia itself was named as a middle ground, the land between two great rivers – the Tigris and the Euphrates. At a similar geographical scale, and as we will see later in Chapter 7, Finnish identity is poised at the middle ground between the east and the west of Europe (Paasi, 1991). A point of significant academic interest centres on the extent to which the east–west and north–south orientations of urbanisation processes at these different geographical scales actually compound or negate each other.

Moreover, the metaphor of the middle ground is neutral with respect to the origins of what might be considered the high points of urbanisation. A conceit of many commentators on, residents and would be residents of, Los Angeles may be that it is indeed the contemporary frontier of civilisation, but it is clear that even within the US this is not true. Although an urban sprawl, Los Angeles is associated with the distinct multiple centres of edge cities. However, the spectre of 'edgeless' cities – edge cities without the sense of place – has been raised in one recent study (Lang, 2003). Here the urban frontier appears to have shifted eastward. 'Miami is like a chunk of Los Angeles that broke off and found its way to the East coast', except that 'South Florida is the edgeless metropolis incarnate. It is simply the most centerless large region in the nation – Los Angeles minus the focal points' (Lang, 2003: 69). Moreover, as Garreau himself notes, 'Europeans can produce new urban environments that are every bit as hard, sterile and contrived as Americans' (Garreau, 1991: 236).

The metaphor of the middle ground also serves to highlight the degree of homogeneity or isomorphism possible in post-suburban function and form produced from the international circulation of ideas and practices. There has been an internationalisation of the state (Glassman, 1999) visible in important instances such as offshore financial centres and industrial enclaves. Transnational local authority networks might be another example of the way in which, in this case local, state practices have been transferred across national boundaries. Indeed, this is the focus of Chapter 3 where we seek to introduce the case-study post-suburban municipalities which form the basis to this study. Also, however, the very *idea* of a particular 'city' such as Los Angeles is something that has travelled internationally to appear as a synecdoche in local debates regarding suburban growth (Forsyth, 1999) including the example of Getafe on the edge of Madrid which we discuss in Chapter 5. Finally, a degree of isomorphism in the development of post-suburbia may have been promoted, as is the case with urban mega-projects, by the transnationalisation of the planning and architecture professions and the finance, development and construction industries (Olds, 2001; Ward, 2005). It hardly needs identifying, but perhaps the most notable example of the power and pervasiveness of concepts in the urban sphere – albeit one refracted subtly through the lens of different national contexts – is Ebenezer Howard's garden city idea. Moreover, it is apparent that American concepts were grafted onto what has come to be regarded as the most British of planning ideas (Hall, 2002). Including the likes of garden suburbs or cities, gated communities and edge cities, Dick and Rimmer (1998) have therefore argued that there are several elements that are reasonably common to cities in North America and East Asia.

More fundamentally, the metaphorical middle ground draws our attention to the very active processes of urbanisation that we are concerned to study. It draws our attention to the problematic distinction between urban and rural and hence to the processes underlying transitions between the two. Indeed, this is one major lesson to be drawn from research on the extended metropolis in Asia. So for example,

> While 'rural-urban' is often proclaimed as a continuum . . . it is always applied as a dichotomy. . . . In terms of first-order differences, there are meaningful distinctions between primate cities and unsettled wastelands, but moving away from these end points towards the middle, does

urban-rural continue to discriminate the most important differences or identify the most important similarities? (Koppel, 1991: 50)

The middle ground of mixed urban and rural land uses and activities is one that has furnished surprisingly few terminological innovations. It may be that closer inspection reveals similarities between patterns and processes of contemporary urbanisation east and west – a point to which we return in the following chapter. The point to be drawn from this discussion is that once one disaggregates urbanisation in terms of dominant functions, forms and processes associated with major agents, it reveals itself as neither east nor west, or indeed urban or rural. It is the middle ground of the various rural–urban and urban–rural transitions that is the most interesting and perhaps the most out-of-analytical focus.

Once one begins to disaggregate processes of urbanisation in this way, one begins to see partial similarities in the functioning and form of post-suburban areas in Asia, Europe, North America and presumably in Africa and South America that express themselves, not so much as urban isomorphism but as variations on isomorphic themes.

1.2   Post-suburbia: A brief history of forms

The purpose of this section is to identify the continuities and discontinuities that exist among what have been considered to be post-suburban forms of development. What we see here are some important continuities – or an evolution – in the geographical and physical form of post-suburbia alongside important discontinuities in the processes involved in its creation.

'Edge city's problem is history. It has none' (Garreau, 1991: 236). The apparent novelty of some post-suburban forms can obscure fundamental continuities in function. This is partly a product of the fact that 'suburbia [and one might suggest post-suburbia] is the hinge, the connection between past and future' (Hayden, 2003: 245). Suburbia has long been invested with a powerful symbolic and ideological content – one that has evolved over time. In medieval times, suburbs were often the site of marginalised populations and activities placed outside of city proper. However, as the urban economy grew, the notion of suburbs as a marriage of town and country arose (Harris and Larkham, 1999: 4) so that the last two centuries of suburban development have seen these places invested with positive connotations that are now a significant driving force in the momentum of urban sprawl. Even within this time frame, Hayden (2003) is able to identify seven different vernacular patterns of suburbia and post-suburbia involving distinctive development practices, building techniques, marketing strategies, architectural preferences and the like. Furthermore, within this time frame, suburbs have been invested with different meanings in different national contexts.

There are several 'myths' that frequently have obscured important continuities and discontinuities in urban development at the edge (Harris and Larkham, 1999). 'One such is that suburbs have frequently been assumed to be socially homogenous. A second such myth that has endured is that they are uniform in form or appearance.' Harris goes on to note that 'different types of suburbs were made in different ways. The implication is that they may be distinguished not only by their form and social character but also by their manner of development' (Harris, 1999: 92). Here Harris, Larkham and Lewis are concerned, along with others, to highlight the role of and articulation between different agents – developers, land owners, governments, residents – in the process of urban development. All of this leads Harris and Larkham to highlight the existence of distinctly different types of suburbs. And in this respect, Lewis (1999) has identified four types of industrial suburb alone.

Though, as we have seen, not all suburbs could be considered residential in character, it is the acceleration of employment decentralisation that Fishman saw as undermining the traditional meaning of suburbia. Instead, then, 'with the rise of technoburb, the history of suburbia comes to an end' (Fishman, 1987: 17). Specifically, technoburbs represented new dispersed concentrations of employment, with some housing, activities at the edges of major cities. Subsequently, the term 'edge city' has gained greater notoriety as a means of defining post-suburban forms and processes. The journalist Joel Garreau (1991) first used the term 'edge cities' to capture the essence of these new employment centres. He defined these edge cities in terms of five criteria, and on the basis of these he identified nearly 200 edge cities existing or coming into existence in the US alone.[1]

More than some other pieces of terminology, edge city appears to have attracted criticism for overlooking the longer lineage of urban development at the fringe of major cities. 'Every few decades someone discovered, yet again, that urban patterns in American cities have never been as tidy as the diagrams produced by reformers, social scientists, and planners would suggest' rather 'the historical record suggests... Edge cities are not a new phenomenon. We can trace conceptual roots back to Ebenezer Howard's garden city and the planned dispersion of the industrial city' (Hise, 1997: 215). Hise's comments are intriguing in their juxtaposing two ideas – edge city and garden city – that, at first glance at least, appear antithetical. Howard's ideas were powerful and pervasive within the planning and architecture professions. But, as Hise's comment suggests and as the account of Espoo in Finland provided in Chapter 7 of this book illustrates, the garden city idea has had markedly different generative effects in terms of urban forms in different national settings.

Fishman's (1987) descriptions of technoburb highlight its incomprehensibility and how its lack of clear boundaries defies incorporation into 'tidy diagrams' of urban form. What we have here is altogether more diffuse forms of urbanisation that have been taken as the signature of what Lang (2003) terms 'edgeless cities'. Most recently, then, the mantle of the ultimate expression of urbanisation processes has been claimed for what have been termed 'edgeless cities' (Lang, 2003). It has been estimated that they contain as much as two-thirds of all the office space outside of downtown areas. They are 'the unmarked phenomenon of the new metropolis' (Lang, 2003: 5) and have been described as illimitable 'cities in function . . . but not in form, because they are scattered, unlike traditional and even some suburban office development' and as such 'not even easy to locate' (Lang, 2003: 1–2).

What can we take from this brief review of suburban and post-suburban forms and processes of development? There do indeed appear to be some continuities or variations on themes. For a start, post-suburban forms continue to be invested with political or ideological meaning carried over from suburban forms. For Fishman, 'suburbia is more than a collection of residential buildings: it expresses values . . . deeply embedded in bourgeois culture' (Fishman, 1987: 4). And although Fishman sees in technoburbs or edge cities a distinct break from these ideological overtones, it is clear that despite their primary economic function, edge cities are also significant lightning conductors for locally generated ideology and politics (Garreau, 1991). How then do we reconcile this with the view that 'In general it is the political meaning of "the suburbs" which is most distinctive from one country to another'? (Harris and Larkham, 1999: 12). The answer is that urban development at the edge of major cities continues to be invested with a reasonably common ideological content, albeit one refracted through different national institutional and political settings. As Teaford has neatly observed, 'basic to the emerging post-suburban polity is the tension between suburban ideals and post-suburban realities. . . . Economically these areas may have become post-suburban but intellectually and emotionally they were solidly suburban' (1997: 5).

Second there is a sense in which the increasingly diffuse forms of urbanisation taking shape defy description. But as Harris and Larkham note, some definitions of suburbs – notably those emanating from and pertaining to Britain – have been quite fluid about their physical location in relation to fixed city boundaries. This is an important point and one that has in subsequent elaborations of alternative suburban and post-suburban forms been the focus of much urban theory. There is then a degree of continuity apparent across what are regarded as different suburban and post-suburban forms over time. As early as 1925 and associated with the Chicago School paradigm of the singular self-contained monocentric city, Louis Wirth had identified the growth and subsequent submerging of satellite urban developments – a process of which we have been reminded recently by Walker and Lewis (2001).

The theme was one that Jean Gottman returned to a few decades ago when outlining the emergence of megalopolitan development of the eastern seaboard of the US, albeit primarily with respect to manufacturing industry let alone tertiary activities. Gottman described how 'On the whole, suburbanization of manufactures proceeded apace in various directions around old central cities and between them, often turning "exurbia" into more ordinary "suburbia" and constantly expanding the more densely occupied and trafficked districts of megalopolis' (Gottman, 1961: 483). Gottman's descriptions of the development of megalopolis in some senses prefigure the most modern of post-suburban forms – Lang's edgeless cities – and make such continuities in forms readily apparent.

1.3   Centre, edge, hinterland

Almost by definition the study of processes of urbanisation at the edge of major cities is conducted in relation to established geographical boundaries. As such there is a sense in which 'suburbia can never be understood solely in its own terms. It must always be defined in relation to its rejected opposite: the metropolis' (Fishman, 1987: 27). The ideological and political content of post-suburban development has been defined in relation to something considered as the city proper – whether broadly 'good' or 'bad' – since at least medieval times. This particular relationship has become more complex over time.

The problem of the relationship between cities and their suburbs and surrounding areas is by no means new. At least since the major expansion of European urban areas began under the influence of the industrial revolution, there has been in general a discrepancy between the administrative boundaries of cities on the one hand and the real extent of the agglomeration on the other. (Council of Europe, 1993: 5)

This geographic dimension forms an enduring facet of the identity of post-suburbia and its political and ideological basis.

The literature on the extended metropolises of East Asia adds an important middle-ground category – *desakota* – to these familiar urban–suburban relations. The term *desakota* has been composed from the Indonesian words *desa*, meaning village, and *kota*, meaning town, to denote vast areas of high population density associated with mixed urban and rural activities (McGee, 1991). As a result, this literature alerts us to a wider set of relationships between central cities and their *desakota* edges on the one hand and further afield hinterlands on the other.

This literature also suggests that mixed uses – residential, industrial, agricultural, retail and idle land – is a characteristic of post-suburbia in need of further examination of itself and in comparative perspective. In fact, explicit mention was made of the mixing of uses – albeit on a large scale – as early as the 1960s in Gottman's (1961) description of megalopolitan growth on the eastern seaboard of the US. Similar megalopolitan growth can be detected in some East Asian countries where the in-between areas are more intensely populated and cultivated. Commenting on subsequent and finer-grained patterns of urbanisation Fishman (1987) also drew attention to the mixing of activities in emerging technoburbs. And this leads us to speculate whether such mixing might – albeit at variable geographical scales in different settings – be a more common feature of post-suburbia in all settings than may hitherto have been entertained.

Such concerns have been overtaken in recent work which refuses to locate post-suburban processes of development in relation to established monocentric models of the city. For Soja, 'what was once central is becoming peripheral and what was the periphery is becoming increasingly central . . . with the intensive urbanisation of the suburbs into outer cities or edge cities' (2000: 152). As such, what may be needed is, as Hise (1997: 12) notes, 'an elastic interpretative framework'.

However, there is a more fundamental sense in which Soja sees 'a significant transition if not transformation taking place in what we familiarly describe as the modern metropolis' (2000: xii). Here Soja is concerned to re-imagine the urban through what he describes as 'thirdspace' analysis. Viewed through the lens of thirdspace 'the spatial specificity of urbanism is investigated as fully lived space, a simultaneously real-and-imagined, actual and virtual, locus of structured individual and collective experience and agency' (Soja, 2000: 11).[2] Amin and Thrift's (2002) subsequent attempt to remove urban studies from the straightjacket of fixed reference points of bounded territories and to offer up a relational geographic perspective on the urban could be considered to provide the sort of conceptual apparatus needed to develop such a 'thirdspace' interpretation of contemporary urbanisation. They prefer to concentrate on the fluid geographies associated with multiple notions of community.[3]

In this book we choose not to embrace Soja's 'trialectics' or Amin and Thrift's 'ecology of circumstance' as analytical tools. Nevertheless, the purpose of Chapter 2 is to contribute to a vocabulary to capture post-suburban forms. It does this by concentrating more particularly on the sorts of processes and agents underlying the development of post-suburban Europe.

1.4 Placing post-suburbia

Bodies of literature have grown up around what have been regarded as distinct forms such as suburbs and edge cities but these have rarely been incorporated into, or captured the imagination within, broader analyses of urbanisation. It has been argued that, until recently, the field of urban studies has been 'underspatialised' (Soja, 2000). Recently, however, several intersecting academic research traditions have begun to redress this situation. Interest in world or global city formation and competition has occupied a leading position within the field of urban studies (Friedmann and Wolfe, 1982; Sassen, 1994; Taylor, 2004). To this can be added more specific interest in the new urban politics (Harding, 1997; Imrie and Raco, 1999; Le Galès, 2002) and more prosaic concerns with the renewed possibilities for and practicalities of metropolitan government (Herrschell and Newman, 2003; Lefevre, 1998; Newman, 2000).

These research agendas speak to, and can be incorporated within, a broad-based concern with the contemporary rescaling of socio-economic processes and state practices (Brenner, 1998; Smith, 1992; Swyngedouw, 1997). Taking these bodies of literature together, the emphasis has been upon exploring the rise of nascent sub-national institutions, politics and territories, defined variously at the scale of cities, city-regions and regions. While some of the aforementioned literature touches upon post-suburban developments, the analysis of these moments of contemporary urbanisation has been left largely outside of these leading debates.

Despite their importance as centres of economic activity and population, and despite some notable academic concern with post-suburbia

(Beauregard, 1995; Keil, 1994; Keil and Ronneberger, 1994), these developments have attracted comparatively little attention in comparison to, for example, the academic interest in global cities. Where post-suburban forms have attracted attention they have tended to be subsumed within these discussions. Charlesworth and Cochrane (1994), for example, identify the importance of post-suburban developments within processes of regionalisation operating in the south-east of England. Post-suburban developments have also attracted some attention within discussion of the functioning and rational government of cities and city-regions. Here, their significance has been viewed in terms of their contribution to the entrepreneurial efforts of established city governments to 'enlarge their spaces of engagement' (Cox, 1998) and to compete internationally for resources and investment (Lefevre, 1998).

The purpose of this book is to focus on urban development at the edge of major cities and to bring it centre stage in discussions of contemporary processes of urbanisation. We want to highlight the often formative contribution that such places and the agents of development thereof play within the development of the wider fabric of cities, city-regions or regionalised urban systems. As Hayden (2003: 11) suggests, 'in the spaces of the suburban city lie metropolitan complexities'. Our purpose in examining the function and form of post-suburbia in Europe is also to illustrate the parallels that exist with urban development at the edge of major cities elsewhere in the world. Drawing upon insights from east and west, in Chapter 2 we attempt to piece together a broader conception of the processes of growth, functioning and form of post-suburbia. Yet this broader view of post-suburban developments needed to capture the lineage and complexity of post-suburban settlements in Europe surely in turn can add something to the analysis of such forms elsewhere – especially so in the US context. Comparative analysis of post-suburban developments has been blighted by an over-concentration on physical *form* which may well be the most exceptional aspect of its development in different national settings. But beneath differences in the appearance of post-suburban forms in different national settings from North America to Europe to East Asia are partial commonalities in the function of and the processes by which these places have grown.

1.5   Structure of the book

In the following two chapters we establish the theoretical and policy context for the examples of post-suburban growth presented in this book. This book draws upon research carried out in five among a unique ten strong European network of 'self-styled' edge cities. To be sure, these places do not conform to the definition of an American edge city. Yet, following the sort of argument developed above and in respect of the broad thematic framework developed in Chapter 2, these places do display variations on some reasonably common post-suburban themes.

In Chapter 2 we develop the broad thematic framework used to discuss the empirical material contained in subsequent chapters. First, we make a distinction between those spaces that are administratively created or planned to a large degree by the state and those which are functionally dynamic. Second, and following on from this, we highlight the artificiality of this distinction when drawing attention to a range of different public- and private-sector agents involved in the processes by which post-suburbia is being constructed. Third, we draw attention to the potential for the mix of functions in post-suburban areas to evolve over time. In the light of the recent interest in technoburbs and edge cities, the function of post-suburbia is often seen as primarily economic. However, we argue that these places are also socially and politically dynamic and, of course, that the function of any given post-suburban settlement may change over time. Fourth, we consider the various ways of interpreting the geographical form of post-suburbia. An important purpose of this review is to highlight that an emphasis on the *form* of post-suburban areas may well obscure points of genuine commonality in their *functioning* and the *processes* by which they have developed.

Chapter 3 then introduces the European edge cities network. It considers the five case-study post-suburban municipalities and discusses the sense in which these places have defined themselves as edge cities. We go on to discuss the extent, nature and benefits of networking and policy transfer among the municipalities noting how the network itself has formed a part of the attempts of local governments to fashion a sense of place identity.

Chapters 4–8 are essentially stories of urban development in five municipalities at the edge of Europe's capital cities. They have been chosen partly with a view to exploring north–south contrasts within Europe with respect to the development of cities (Leontidou, 1990); administrative structures and leadership roles (Klausen and Magnier, 1998a); and styles of, and agents driving, development (Berry and McGreal, 1995).[4]

The cases of Kifissia and Getafe in Chapters 4 and 5 represent patterns and processes of post-suburban development in the south of Europe. We begin our empirical exploration of post-suburban Europe, in Chapter 4, with the case of Kifissia, a suburb of the ancient city of Athens. Of our five cases, Kifissia is the least like an edge city in form or function. Yet there is a value in beginning with Kifissia, as it is the ancient Greek philosopher Plato whose 'theory of forms' provides a heuristic analytical device for appreciating the sense in which there are variations on a theme of post-suburban development. Kifissia itself has a long and distinguished architectural and cultural history as a suburban resort for wealthy Athenians. More recently it grew into a residential suburb before recent leisure and retail developments have transformed it again into a more fully post-suburban municipality. Moreover, distinctively southern European patterns of development have produced – albeit in a different, economically marginal and miniature fashion – patterns of mixed land uses that have been most closely associated with East Asian *desakota*-style urban development.

Madrid was an early invention as the capital of the Spanish nation. Like its near and grander neighbour, Getafe was recently invented by the state as a peripheral employment and residential *space* but has grown to become a *place* with its own powerful sense of identity and actors able to wield influence within wider metropolitan and national spheres. Getafe grew from the 1960s into a dormitory and manufacturing suburb to the south of Madrid. Part of Madrid's southern 'red belt', it is associated with the grass-roots political movements of the 1960s and the 1970s whose legacy has conferred a lasting political capacity within the municipality which has been reactivated recently in the pragmatic mayoral politics of Pedro Castro.

Noisy-le-Grand at the eastern edge of Paris is perhaps the most unique of our chosen post-suburban areas. As we discuss in Chapter 6, Noisy remains the product of state strategies of employment and population decentralisation as part of a new town growth-pole. In form very different from contemporary post-suburban developments in North America, it nevertheless suffers from the sort of placelessness commonly associated with North American urbanisation (Augé, 1995; Kunstler, 1993). It has yet to outgrow problems associated with its being created as a state space.

Our final two northern European cases actually represent the closest approximations to US-style post-suburban development but perhaps in different respects. Chapter 7 describes how Espoo has grown rapidly to become the second largest city in Finland and actually approximates along key dimensions to North American-style urbanisation. The rapid growth of this city represents a uniquely Finnish adaptation of Ebenezer Howard's garden city ideals. Somewhat paradoxically, despite its origins and development within the Nordic welfare-state system, it presents an example, to place alongside Hise's discussions of the planned nature of Los Angeles' growth via the transference and mutation of an internationally influential ideology. In this respect, Espoo comes close to the growth machines that are typical in many North American cities (Logan and Molotch, 1987; Molotch, 1976).

Finally in Chapter 8 we discuss the growth of Croydon from a market town to a dormitory suburb to a significant suburban office and retail centre within the London and south-east of England. It has none of the appearance of a North American edge city but yet within the European setting its emergence as a major employment centre means that it has performed a similar function to an edge city, and has grown as the product of the sort of urban regime politics so apparent, in North America (Elkin, 1987; Stone, 1989).

In a concluding chapter, we return to the issues discussed in the opening two chapters with the aim of highlighting the partial similarities in form and process in each of these five places and we suggest further afield in North America and East Asia. Post-suburbia is revealed as variable and friable. Various agents – such as land owners, developers, populations and governments – are seen to have alternate roles in shaping the different examples of post-suburban Europe. Finally, since urban sprawl is a topic of not only intense academic debate but also policy interest, the chapter considers how we might learn from the various experiences of contemporary processes of urban development at the edge of major cities.

2
Closer to the Edge: Function and Form in Post-Suburban Europe

> *Edge City* . . . Charlie closed his eyes and wished he'd never heard
> of the damn term.
>
> Tom Wolfe, *A Man in Full*

2.1 Introduction

The term 'edge city' (Garreau, 1991) is something that academics, along
with Tom Wolfe's developer hero Charlie Croker in the opening quotation, have come to use with no little anxiety. As Soja notes 'for much
of the world, the Edge City maxim, that every American city is growing
in the fashion of Los Angeles has become much more of a foreboding
than a hopeful promise' (2000: 401). Moreover, while the term 'edge
city' takes its place in a welter of terminology deployed to help chart
the complexity of modern forms of urbanisation, its precise relevance in
the European setting is highly questionable (Ghent Urban Studies Team,
1999; Lambert *et al.*, n.d.). It will come as no surprise, then, that we avoid
the term 'edge city' or any explicit attempt to define the 'European edge
city'. And as we will see in the next chapter, no definition of an edge
city in the European setting has been forthcoming from the European
network of self-styled edge cities. Rather, in keeping with the diversity of
experiences of urbanisation in Europe, and in keeping with the diverse
empirical cases reported later in this chapter, we use the term 'post-
suburbia' to capture a profusion of terminology relating to a nascent
urban form and over which there is only partial consensus.

In what follows, we first develop broad themes relating to the
contemporary rescaling of functional processes and state practices and
structures. These themes permit us implicitly to begin to distinguish

European post-suburbia from its North American counterpart, by considering the alternate agents, fungible functions and friable forms involved.

2.2 Functionally dynamic or administratively created post-suburbia?

The recent and sizeable academic interest in the rescaling of political, social and economic processes derives from a desire to understand the repositioning of different territorial scales within an ever more integrated world economy. As Brenner describes, 'the post-1970s wave of globalisation has significantly decentred the role of the national scale as a self-enclosed container of socio-economic relations while simultaneously intensifying the importance of both sub- and supranational forms of territorial organisation' (1999: 435). Terms such as 'multilevel governance' (Marks *et al.*, 1996), 'glocalisation' (Swyngedouw, 1997) and the 'relativisation of scale' (Jessop, 1999) have been used to capture the interconnections between processes at various spatial scales. In particular, interest has centred on the renewed potential of sub-national regions and cities – when set against nations – in such multiple levels of governance. So, for example, renewed possibilities for political and governmental mobilisation in the regions have been the subject of quite intense study across Europe (Keating, 1997; 1999). Similarly, some of the contours of an emerging Europe of the city-regions have been charted (Harding, 1997; Le Galès, 1998). Yet some scales have remained invisible to much of this research effort. Notable in this respect is the sub-metropolitan scale represented by post-suburbia. We therefore address the urgent need to consider the position of post-suburban forms within the contemporary rescaling of socio-economic processes.

Despite their importance as centres of economic activity and population, and despite some notable academic concern (Beauregard, 1995; Keil, 1994; Keil and Ronneberger, 1994), post-suburban developments have, of themselves, attracted little attention. Instead, following an earlier interest in global cities, the primary academic interest has focused on major cities as a window onto the possible emergence of a Europe of city-regions. Furthermore, planning strategies for these cities have concentrated on central city areas that embody 'world city spaces' (Newman and Thornley, 2005: 275) rather than on their entire territory, let alone their periphery. On the one hand, then, post-suburban forms have often been subsumed within discussion of broader sub-national regions. Charlesworth and Cochrane (1994), for instance, see the significance of post-suburban formations in terms of processes of regionalisation.

> This underplaying of the regional dimension is particularly problematic in the light of developments which point towards the extensive networks of 'suburbs' or 'edge cities' and the emergence of what have been called non-places, each of which nevertheless has its own institutions of local governance and networks of local politics. (1994: 1725)

On the other hand, where post-suburban areas have been incorporated specifically into analysis of cities and city-regions, their role has been defined in terms secondary to that of central city areas. The significance of post-suburban developments has been viewed in terms of their contribution to the entrepreneurial efforts of major city governments as they seek to 'enlarge their spaces of engagement' (Cox, 1998) to compete internationally. In this book we want to place post-suburban populations and institutions centre stage, highlighting their importance to contemporary processes of urbanisation.

The object of analysis within the literature on the rescaling of socio-economic processes is upon the study of process and not pre-established administrative territories (Brenner, 1999; Jonas, 1994; Swyngedouw, 1997). If in theory the object of analysis is 'the study of process through which particular scales become (re)constituted' (Swyngedouw, 1997: 141) in practice, because of the continual rescaling or geographical fluidity of processes, there is a necessity to analyse both process and pre-existing scales. There is an irony here as, although there is a very real difficulty in speaking of post-suburbia as given, there is nevertheless a need to suspend one's dissatisfaction with such an idea in order to appreciate the way in which such places are socially constructed. In this respect then, there are at least two analytical devices that might be deployed in order to capture this near constant re-scaling of processes. One such device involves the refocusing of analytical attention upon boundaries or boundary regions as the objects of analysis (Paasi, 1991, 2000). Here the object of analysis would be a region defined so as to straddle existing administrative boundaries. This represents an idiographic approach but one which is defined in order to problematise and disrupt the coherence of established territorial boundaries. Notable in this respect then is the intense policy and academic interest in European cross-border regions (Perkmann, 2003).

A second device involves making a somewhat artificial distinction between the relative geographical fixity of administrative or state practices, structures and agents on the one hand and the relative geographical mobility of functional (non-state) economic, social and informal political processes and associated agents on the other hand. Actually, as Brenner (2002) points out, practical questions of urban administrative reforms in the US have long been conducted in terms of this distinction. This is also reflected in the work of Paasi, who conceptualises the emergence of regions in terms of a number of stages (Paasi, 1991: 243). Paasi makes an implicit distinction between functional economic, social and political processes and spaces on the one hand and administrative or state processes and spaces on the other, suggesting a progression from the initial role of functional economic, social and political processes in shaping territory towards the crystallisation of that territory in institutions and presumably administrative or state structures. A similar explicit distinction is made by, among others, Bennett (1997) and Keating (1997), with Bennett arguing that administrative structures tend to lag behind or shadow functional processes and as a result there is constant 'under-bounding' of state spaces.

The significance of post-suburban developments is self-evident when deploying the first of these analytical devices. As the subtitle of Garreau's (1991) book suggests, the urban edge represents a boundary or frontier region. Practices of internal integration and external differentiation are central to the creation of territorial boundaries (Paasi, 1996) and are visible in the actions of agents constructing post-suburbia in different contexts. In the remainder of this chapter we also wish to deploy the second of these analytical devices. By utilising these devices we wish to make an initial contribution to charting post-suburbia in Europe and contrasting it with its North American edge city corollaries.

The distinction between the fluidity of functional economic, social and informal political processes on the one hand and the relative fixity of state structures and practices on the other might be regarded as a 'first cut' heuristic device for analysing the dynamics of post-suburban Europe. Yet it is a distinction that clearly is of wider relevance to appreciating, for example, the development of diffuse urban forms – including edge city style elements – in East Asia that actively have been planned by developmental national states.

Altered states? Post-suburbia – from problem container to growth pole

It is commonly assumed from the North American literature that post-suburban developments are the product largely of market forces surrounding the dynamics of land, commercial and residential property markets. This is the popular assumption behind the concept of edge cities whose function and dynamism precedes the formation of 'shadow' governmental structures. As Garreau notes 'edge cities . . . seldom match political boundaries' (1991: 185). There is also a sense in which some settlements at the edge of major cities in East Asia have developed without significant state involvement since 'desakota zones are to some extent "invisible" or "grey" zones from the viewpoint of the state author-ities. Urban regulations may not apply in these "rural areas", and it is difficult for the state to enforce them despite the rapidly changing economic structure of the regions' (McGee, 1991: 17).

Yet, there is no reason why this relationship between functional processes and spaces and administrative processes and spaces may not operate in reverse, not least because of the strategic and spatial selectivity of the state (Gottdiener, 2002; Jessop, 1990; Jones, 1997; Walker, 1981). Indeed the intensified rescaling of the state seen by Brenner and others presents this very possibility. Here 'rescaled state institutions are increas-ingly viewed as a central means of delineating locally and regionally specific growth poles through which capitalist territorial organisation can be mobilized "endogenously" as a force of production in the world market' (Brenner, 1999: 476).

The contemporary rescaling of the state has implications for sub-national regions. Keating (1997), for example, describes how the rise of regions within many European nations can typically be characterised in terms of a series of central government–led administrative settlements. Brenner, however, draws attention to city-regions as perhaps *the* scale around which state structures and practices are being redefined most intensively. He highlights the centrality of local elites and economic development objectives as the drivers of this rescaling of the state.

> The city-region is being mobilized as the key institutional pivot between an internal realm of co-operation, administrative co-ordination, embedded firms and socio-spatial solidarity and an external space of aggressive territorial competition, intergovern-mental austerity, mobile capital flows and unfettered market rela-tions. (Brenner, 2003: 310–311)

All of this may be placing too heavy a burden upon what appear to be, by Brenner's own admission weak, often contradictory and open-ended processes. So, for example, Keil highlights the indeterminacy of neoliberal municipal ideologies in resolving problems of metropol-itan governance, alternately promoting amalgamation in Toronto and secession in Los Angeles. Certainly, the frailties of metropolitan planning agendas repeatedly have been exposed by resistance from suburban municipalities. Moreover, and of specific interest to our purposes here, it may be as well to entertain Teaford's (1997) idea that post-suburbia represented, at least for a time in the 1970s and 1980s, a middle way compromise with the growth of county-level government in the US.

Moreover, the coherence and meaning of such state rescaling clearly varies across Europe let alone further afield. Newman (2000), for instance, notes how, in France, regional reforms, including joint planning between the central state and regions, failed to invigorate regionalism because of domination by the central state. In the European context then the idea of 'strong states and "dependent" cities appears, as yet, to have plenty of mileage left in it' (Harding, 1997: 296). This, we can suggest, will also be the case in the formation and development of post-suburban areas in the European case, in the sense that state institutions, especially non-local ones, will have an important bearing on the ability of agents to construct distinct post-suburban identities.

Another ingredient of the contemporary rescaling of the state has been a loss of some capacities of the nation state to supranational governmental bodies such as the European Commission. The European Commission has been active in the establishment of numerous inter-urban networks as a means of delivering financial resources to promote social cohesion. Notwithstanding the fact that these networks are heavily imbued with and underline neoliberal principles of inter-urban competition, those that have been established also embody 'new political spaces for cities to challenge extant state structures and relations' (Leitner and Sheppard, 2002: 514). And, as we will see in the following chapter, such networks have indeed been harnessed by post-suburban municipalities as they attempt to widen their spheres of influence within metropolitan and national state machineries.

The point to this discussion is simply to highlight that this intense period of rescaling of administrative or state practices and structures may actually set in train a rescaling of functional economic, social and political processes. This can manifest itself in direct and comprehensive interventions by the state such as the creation of new towns or less comprehensively in terms of the restructuring of local government boundaries and responsibilities. There are two different outcomes that we can consider here. The first, and perhaps more common, outcome is where central national state strategies have created 'nowhere' places. This is a familiar scenario in which 'states impose spaces on places' (Scott, 1998; Taylor, 1999: 14). One unintended consequence of comprehensive state intervention and planning has been the production of numerous suburban settlements that amount to little more than 'problem containers'. In fact, the history of development at the edge of major cities is replete with unintended consequences so that 'the origins of contemporary sprawl, as paradoxical as it might seem... are deeply rooted in the long-standing concern of urban reformers over excessive density' (Bruegmann, 2005: 449). The case of Noisy-le-Grand, presented in Chapter 6, provides a striking instance of post-suburbia as a state-created nowhere. Here then it is worth remembering that, historically speaking, 'the symbolic importance of the modern ideals of integration and cohesion was... radically different from their effects in practice. Beneath the universalising rhetoric, modernising cities were always about rupture, contradiction and inequality' (Graham and Marvin, 2000: 42).

The second outcome is where state restructuring actually galvanises social, economic and cultural processes to create meaningful places. 'Although initially imposed, boundaries can... become embedded in society and have their own effects on the reproduction of material life. In this way what were spaces are converted into places' (Taylor, 1999: 14). Modernist ideals of the planned city, and by implication post-suburban settlements, may turn out to have had, when seen in the long-run history of city building, a very short tenure indeed. Nevertheless the generative effects of these interventions, as the discussion above highlights, may live on in economically, socially and politically vibrant post-suburban places.

Both of these outcomes are important to consider in the light of the strong states of European nations where territorial redefinition has tended to be led by administrative reform such as devolution. However, irrespective of which outcome may predominate in any particular setting, it is somewhat of a misnomer that edge cities should be associated with 'spontaneous' urban development. As Lang notes, 'ironically, Garreau's edge city criteria... originated in a planning document drafted because the market was not generating edge cities' (Lang, 2003: 99) with planners in Phoenix concerned to create a polycentric urban structure from what they feared was turning into a centreless sprawl of office developments. Further, as Wolch *et al.* (2004: 2) have recently reiterated regarding the edge cities of Southern California, such a view 'ignores the fact that these areas have grown around and depended entirely on public-funded highways, and, in some cases, airports and government facilities... They have been influenced by federal and state policies, including mortgage subsidy programmes, highway building programmes and tax systems.'

2.3 Alternate agents: The producers of post-suburbia

The distinction between functional processes and the scales at which these operate on the one hand, and administrative or state practices and structures on the other hand, is also important in connection with the suggestion that the focus of urban politics has shifted from social welfare policies towards economic development objectives (Mayer, 1995) and from urban managerialism towards urban entrepreneurialism (Harvey, 1989). We have already discussed the role of the state as prime architect of some post-suburban settlements. Indeed it is the very discrepancy between the plans for and the reality of our urban landscapes that directs attention to the range of agents involved in their production (Ambrose, 1994; Whitehand and Carr, 2001). In what follows, we discuss a number of ways in which the growth of post-suburban areas could be considered to have been planned in more subtle senses – as the product of some combination of agents. In doing so, the artificiality of distinctions between what we commonly regard as public-sector planning and spontaneous free-market forces is revealed.

Privatised planning and corporate post-suburbia

A degree of planning or intentionality can be detected behind even the most apparently spontaneous of North American urban forms. Indeed, the promotion of suburban development was integral to the expansion of American capitalism (Walker, 1981), constituting a 'suburban industrial complex' (Rome, 2001) in which the mass consumption of housing and also household durable goods – including the car – drove economic growth and employment. As such, suburban and post-suburban development has involved and continues to involve

> large scale planning and resource management by private builders, real estate developers and banks. These planned actions are tailored to the various regulations of the extensive planning apparatus existing at all levels of government, and are also supported by federal programs and subsidies... (Gottdiener, 1977: 93)

Gottdiener's (1977) work related to suburban expansion on the eastern seaboard of the US. Echoing this, Hise has described how suburban and post-suburban development in Los Angeles, often assumed to be the ultimate expression of an absence of planning, was in fact underpinned by a good measure of micro-planning albeit one that was not integrated with wider strategic planning visions. One might be surprised to find that Ebenezer Howard's ideas for garden cities were translated into a specifically North American concept of community and neighbourhood planning in which the modernisation of the housing industry and efficient housing production was a core concern.

> The neighbourhood served as a middle common ground. Decenterists and pragmatists conceived it as an integrated whole, tightly segregated within the overall city or region, with its own internal hierarchy from public to private, from the workplace to the individual dwelling.... Cities and regions were less diagrammatic and coherent... Well-planned neighbourhoods became islands of rational planning in a pragmatists' sea. (Hise, 1997: 52)

What were essentially time-limited and strictly geographically contained 'corporate' experiments in Britain became the routinely reproduced pattern of development rolled out across Los Angeles and elsewhere in the US. As such, the 'garden city' idea specifically as well as broader notions of 'garden suburbs' have had a very wide-reaching influence upon urban form indeed (Hall, 2002; Whitehand and Carr, 2001: 183).

Hise's account of the growth of Los Angeles is part of a broader body of work whose contribution has been to offer an alternative to the predominant view of residential-led sub-urbanisation. Instead Hise, Lewis and perhaps Fishman and Garreau are part of a corpus of authors to draw attention to the leading role of employment in suburban and post-suburban developments. Within such developments employers have often, though by no means in all cases, engaged in planning broader communities. These contributions draw attention to the close articulation among agents of the real estate business – such as land owners, developers and builders and the way in which this does or does not in turn articulate with broader land-use and spatial planning practices. Thus, even where larger scale strategic planning has structured developments, sprawl has also resulted – pointing to systemic difficulties in controlling urban sprawl in the US. Here we come close to the boundary at which we find intentionality and planning on the one hand and the private sector on the other. It is a moot point as to whether one describes this as planning or privatism. 'Nothing in Los Angeles demonstrated the tenacity of private developers as convincingly as the course of the planning system' (Fogelson, quoted in Dear, 1996: 92).

Hogan (2003) argues that the case of suburban sprawl in San Diego is more instructive than the archetypal case of Los Angeles, since, paradoxically, it is the very success of planning that has driven sprawl. The problem being that San Diego's 'big picture' strategic planning became a legal-rational and corporate process dominated by big business and big environmentalism. Kunstler interprets these unintended consequences of planning as a case of the capture of public-sector regulatory activity by private-sector interests such as realtors, engineers and so on. What this body of work suggests is that 'the diffuse city gives an "unplanned" impression, but it has arisen out of innumerable individual and – considered on their own – rational decisions' (Sieverts, 2003: 3). Since there appears to be little difference in the actions of public- and private-sector agents in the suburban and post-suburban development, 'decisions made by the planners, speculators and housing developers lead to the same land-use pattern as would result from no planning' (Gottdiener, 1977: 111).

Planning as partnership: Post-suburbia as growth machine or urban regime

The implications of the initial, first-cut, distinction drawn in Section 2.2 need to be pursued in relation to a trend that Mayer (1995) refers to as an expansion of the space of local political action. Here the term 'governance' has been used to describe the way in which a range of private and quasi-autonomous non-governmental organisations have taken their place alongside the local state in local political processes. The term has become axiomatic despite its limitations (Imrie and Raco, 1999) and despite its clear and potentially partial resonance with ideas of growth machines and urban regimes which have grown out of a specifically North American context.

Two influential formulations of urban politics that speak to some kind of public–private partnership in the shaping of urban development are that of the city as growth machine (Logan and Molotch, 1987; Molotch, 1976) and as urban regime (Elkin, 1987; Stone, 1989). Molotch and Logan had in mind the way in which urban economic and physical planning agendas are effectively captured by those private-sector interests that are most dependent on the fortunes of the local economy – ostensibly land-based interests such as real estate companies, property developers and so on. The concept of urban regimes stresses instead the lead taken by the public sector in engaging with private-sector interests in order to shape and achieve urban development objectives.

These theories have been developed in relation to major US cities. And yet 'although it is well-known that the political economy of post-war developments in the US has been predominantly suburban in character... only a handful of researchers have investigated in any great depth the emergence and characteristics of... political regimes in the suburbs' (Althubaity and Jonas, 1998: 150). Indeed, Hayden (2003) places the likes of edge cities within a two-centuries-old lineage of the suburban growth machine. This view might need to be qualified somewhat. Gottdiener's (1977) pioneering work on the politics and planning of urban sprawl on Long Island did indeed suggest that the 'local political party and those businessmen involved in submetropolitan growth tend to merge into something of a land development corporation' (Gottdiener, 1977: 111). Although careful to note growing signs of the accretion of non-residential functions, internal fragmentation and social and racial inequality, Gottdiener's work reflects on the expansion of relatively socially homogeneous residential suburbia prior to its mutation into more fully-fledged post-suburbia. He therefore also made plain that at this time the 'weak' suburban government has had neither the consistency nor the strength of established city growth machines.

The applicability of these theories to the post-suburbia of Los Angeles and Southern California in recent times at least seems clear (Althubaity and Jonas, 1998; Dear, 1996; Jonas, 1999). Dear describes a series of six different intentionalities that have underlain the development of Los Angeles. In doing so, he draws attention to an articulation between the public and the private sectors – growth regimes or coalitions – common in the US.

> The maturation of a distinctly modernist planning in Los Angeles can be seen in the successive emergence of entrepreneurial and state-centered growth regimes at the turn of the century.... It was also a period when... an idealized utopian planning theory was divorced from the localized processes of capitalist urbanization. The most consequential practical manifestation of this fracture was the subordination of the land use planning apparatus to the exigencies of local capital. (Dear, 1996: 97)

Dear is at pains to stress that the minimal sorts of planning enshrined in the format of neighbourhood development have given rise to a lack of intentionality. As he suggests, there is 'no longer a single civic will or a clear collective intentionality behind LA's urbanism' (Dear, 1996: 99).

The specific pattern of development in Los Angeles as it has evolved towards some form of suburban or post-suburban regime or growth machine appears as an extreme instance of a pattern common across the US. The comparatively recent State legislation within the US that has sought to limit urban sprawl since the 1970s whereby 'cities typically used zoning to uphold property values and promote economic development, has not prevented environmental degradation. Worse yet, many counties did not use their regulatory powers at all' (Rome, 2001: 229).

The growth machine and urban regime theories speak to the sort of autonomy of local actors apparent in the US and are, of course, less well disposed to analysing the manner in which the activities of local agents are structured by a system of multilevel governance in Europe (Newman and Thornley, 2005). Rather, the local and central state and the public sector more broadly play a much more important role in economic development strategy at the urban scale (Harding, 1991). Moreover, Jouve and Lefevre (cited in Newman and Thornley, 2005) argue that local political elites in European cities have rarely been able to exert their autonomy from actors operating at higher administrative levels. As such 'the institutions and networks found in promoting redevelopment in European cities simply do not have the local *gravitas* of a growth machine or a regime' (Harding, 1997: 299, original emphasis). By comparison with American public–private partnerships, which frequently vest most power in the private sector, European public–private partnership experiments remain fairly limited, except in the United Kingdom (UK), and city councils still have strong capacities for initiative and control (Le Galès, 2002: 259).

The concepts of local dependence (Cox and Mair, 1988, 1991) and the related idea of 'spaces of engagement' (Cox, 1998) are less prescriptive than growth machine or urban regime theory. Both public- and private-sector actors have a degree of local dependence which has a complex relationship to participation in local political coalitions (Wood, 2004). Land-based private-sector interests may well be heavily dependent upon local economic fortunes, but the likes of manufacturing and service companies are only partially dependent upon the state of the local economy due to their national and international markets, recruitment of labour and so on. While public-sector actors such as local governments are heavily dependent upon the locality (in terms of their political constituency and tax base), they are even not entirely locally dependent due to extra-local sources of revenue and an ability to enlarge their spaces of engagement – deploying political capabilities to gain material and ideological capital. All of the post-suburban municipalities considered in this book are to a greater or lesser degree implicated in strategies to mobilise extra-territorial financial and political resources.

This latter point is of some significance to an appreciation of the dynamism of post-suburbia. Notwithstanding the general observations above regarding the weakness of private–public sector coalitions in urban politics in Europe, a degree of post-suburban entrepreneurialism does appear to be apparent. The entrepreneurialism of post-suburban municipalities appears to have been driven in part by processes of international political and economic integration as these have impacted upon major city-regions. So, for example, the fortunes of central cities and their edges are frequently linked in a city-region consciousness which 'is an important part of the ideology of structural coherence in the region and unites it for the global interregional competition' (Keil and Ronneberger, 1994: 162). The entrepreneurialism of post-suburban areas in conjunction with that of central city areas is a crucial ingredient of the renewal of the metropolitan idea in Western Europe. This is because 'central cities ... are now aware that they need the peripheries in order to develop, or quite simply to keep their place, in the ranks of world cities' (Lefevre, 1998: 22).

Beyond this, however, there are signs of entrepreneurial municipalism independent of these metropolitan agendas. Althubaity and Jonas (1998), for instance, have spoken specifically about the post-suburban municipal entrepreneurship apparent in North America. Post-suburbia is essentially economic in function, but its residents retain suburban imaginaries of local identity (Teaford, 1997: 5). Drawing on Teaford (1997) we can suggest that the new developments emerging at the edge of major cities embody a distinctive set of 'post-suburban' tensions. One key product of this is the potential for post-suburban municipal entrepreneurialism to embody a tension between economic development objectives and constraints imposed by collective consumption expenditures (Althubaity and Jonas, 1998). A second key product of this tension is the potential for post-suburban entrepreneurialism to emerge from secessionary politics (Keil, 2000). Here even the smallest post-suburban municipalities are often able to make heroic appeals to their historical independence from their larger city neighbours. A third key product of post-suburban political tensions is local conflict over growth versus the environment and conservation (Jonas, 1999; Pincetl, 2004) where, given the economic basis of these new post-suburban areas, the former may prevail. Regardless of which interests prevail, the key point here is that a set of organised residential and environmental interests (Rome, 2001) has emerged to make its own contribution to the shape of post-suburbia. However, as the unanticipated effects of growth continue to proliferate, the strength of emerging post-suburban growth machines or regimes may rest in part in what Gottdiener (1977: 167) was able to observe as a splintering into single-issue organised political interests.

It is perhaps less clear whether such post-suburban municipal entre-preneurialism exists in the European setting where, as we have seen, local autonomy is somewhat structured by the central state within what can be viewed as a complex setting of multilevel governance. Some suburban and post-suburban municipalities do appear to have had a strong entrepreneurial stance. Indeed, the key differences between post-suburbia in Europe and North American may centre around municipal entrepreneurialism given that 'state intervention, regional planning schemes and local authorities play a much larger role, with the public sector often acting in a quite entrepreneurial way' (Bontje and Burdach, 2005: 328). To prefigure our own discussion of Croydon in Chapter 8, Saunders's (1983) classic study indicated a long-standing role for busi-ness in local politics while Dowding *et al.* (1999) found that Croydon was unique in their study of eight London boroughs in approximating to a US urban regime.

In sum, various agents such as the state in its various guises, resid-ents, land owners, real estate companies and financial institutions, prop-erty developers and construction companies, major manufacturing and service-sector companies each play a role in the creation of different types of post-suburban settlement. As such, 'an edge city does not simply materialise in the suburban landscape . . . it must first be built. There is . . . a politics to this building process, a politics in which discourse and material reality do not always converge' (Jonas, 1999: 202–221). As a result, 'contestation is the real story of suburbia' as Hayden (2003: 245) has reminded us.

2.4   Fungible functions: The dynamism of post-suburban Europe – economic, social and political

To begin with, the distinction between functional processes and the scales at which these operate on the one hand, and administrative or state practices and structures on the other hand, although somewhat artificial, is important if we are to view the function of post-suburban areas in the round.

Certainly, there is an emphasis on the economic function of post-suburban forms to be found within much of the literature. In part this is indeed appropriate to understanding the increasing significance of post-suburban areas in an increasingly integrated international economy. In the US, studies have documented the superior economic performance of outer suburbs or edge cities over traditional central city areas in terms of jobs growth (Hill and Wolman, 1997). Here, a range of central city–suburban relationships are apparent including not simply suburban dependency but also growing economic independence of post-suburban areas. In the light of this, it is hardly surprising that Keil should describe how 'the new peripheries have become the projection spaces of the emerging global post-fordist economy: the target of investment and accumulation' (Keil, 1994: 134). The economic centrality of urban peripheries is also highlighted by Dear and Flusty who posit the existence of 'a postmodern urban process in which the urban periphery organizes the center within the context of globalizing capitalism' (Dear and Flusty, 1998: 65).

Yet the economic basis of these projection spaces of the post-Fordist economy is poorly understood. Post-suburbia – the likes of edge cities, edgeless cities and inter-urban locations – have rarely been the explicit focus of economic analysis. Following Jacobs (1970) it has often been assumed that suburbs and smaller towns are inert – that they represent collections of unrelated businesses that embody a playing out of 'sterile' divisions of labour. Yet Jacobs (1970) also noted that there has often been little association between physical and environmentally sustainable urban form and the economic success of cities in history. And thus, while suburbs and post-suburban areas may be derided for their lack of sustainability, they may also, by the same token, embody more economically dynamic and innovative urban spaces than hitherto entertained.

Phelps (2004) makes clear that the economy of the urban edge is not best understood in terms of received theories of external economies and agglomeration. Lang (2003) argues that edge cities do not possess unique advantages in terms of place-bound, or technological, external economies. Rather, in some instances their economies embody a combination of different (pecuniary and technological) external economies so that they effectively 'borrow size' (Alonso, 1973) from a larger city or group of cities. All this should not be surprising since Walker and Lewis have noted the origins of many early North American suburbs in initial industrial functions. Similarly, Hise (1997) has documented the manner in which the sprawling expanse of Los Angeles often centred around the lead role taken by major employers. As a result of this, or of the rapidity with which employers have followed residents in the suburbanisation process, Gordon and Richardson (1999) note that the common assumptions regarding the growing commute times and separation of work from residence apparent in the 'Los Angeles school' literature are inaccurate.

Notwithstanding these observations regarding the significance of post-suburban areas in economic terms, in the European (and indeed developing country) context, it would be as well not to overlook the role that the re-scaling of *political* and *social* processes play in the dynamism that post-suburban-based agents and institutions contribute to broader city-regions. These, as Keating (1997) has outlined, are equally significant in devolution in Western Europe. They may also be important at more localised geographical scales including communities at the edge of major cities. One reason for this is that the 'politics of resistance' have been relegated to smaller and more particular scales (Swyngedouw, 1997).

It is clear that much of suburban development in the US has, at least to begin with, been firmly residential in function leading to a particular social and political complexion of such settlements. Indeed, for all the stress placed on the economic role of edge cities Garreau describes how these new urban centres 'are marked not by the penthouses of the old urban rich or the tenements of the old urban poor, but by the celebrated single family home with grass all around' (Garreau, 1991: pre-introduction). The selective outmigration of wealthier and economically active segments of population in North America means that post-suburbia embodies distinctive political tensions. As we saw above, while the likes of edge cities have become post-suburban in terms of their economic function, they remain suburban in terms of the imaginations of their residents (Teaford, 1997).

Even in the US, the social and political complexion of suburbs has become less uniform (Muller, 1981). Outside the North American context, the social characteristics of the population migrating or re-located to the urban edge are less uniform still. Here, the formative role of some post-suburban areas as the origin of significant socialist political movements and of redefinitions of citizenship appears to have been largely overlooked. Holston has argued that areas at the periphery of the São Paulo city-region were significant in redefining notions of citizenship at the city-region scale. 'As in many other metropolises around the world, the urban poor of São Paulo established a space of opposition – the periphery – within the city-region. This space confronts the old culture of citizenship with a new imagination of democratic values' (Holston, 2000: 339). The grass-roots political movements of Spain in the 1960s and 1970s were also significantly those of peripheral urban areas (Castells, 1983) – something we address directly when discussing the case of Getafe in Chapter 5 of this book.

In Europe, the lineage of many post-suburban areas in state redis-tributive and spatial planning policies means that some of these new forms of urbanisation are still materially and discursively imbued with considerable social and political maladies. Yet many post-suburban areas have also outgrown their original identities as containers of social prob-lems and attendant political struggle to become economically dynamic or part of the political mainstream of metropolitan society. 'The peri-phery is not the periphery anymore. In Europe it has ceased to be merely the problem container of cities, perverted product of social reform based on the inner city' (Keil and Ronneberger, 1994: 141). One consequence of this is that many of the social and political movements of the 1970s have conferred a lasting political capacity upon their respective local-ities, being incorporated into mainstream municipal politics and even 'routinized cooperation' with the local and central state by the 1990s (Mayer, 2000: 138).

Hayden's (2003) history of suburbanisation in the US highlights some of the shifts in function and symbolic content of suburbs over time. A final implication of these trends, then, is that it is quite apparent, whether employers have lead or followed, that post-suburbia has been and continues to be multi-functional, and that even the predominant function is subject to change over time.

2.5   Friable forms: Post-suburbia's internal coherence and enlarged spaces of engagement

Finally, an important concept to emerge from the literature on contem-porary rescaling of socio-economic processes and state practices is that of the variable or eccentric nesting of scales (Jessop, 1999; Jonas, 1994; Swyngedouw, 1997). The analytical focus on processes through which scales are socially constructed alerts us to mechanisms of scale trans-formation and transgression through which there is a continually chan-ging or fluid nesting rather than some immutable hierarchy of scales (Swyngedouw, 1997). Post-suburban areas in North America and Europe provide us with an excellent illustration of such a nesting of scales.

Post-suburbia and the new geometry of urbanisation

Speaking from the south-east England context, Charlesworth and Cochrane argue that the growth of suburban and post-suburban areas makes it 'impossible to pretend that local politics are somehow rooted in the experience of free-standing and bounded "localities"' (1994: 1726). Instead, as Keil and Ronneberger argue drawing upon the case of post-suburban developments in Germany, 'core and periphery are not plausible anymore as geometric concepts. Rather, we are dealing with a relational model of spatial relationships manifesting themselves in myriad forms...' (1994: 139).

Here post-suburban agents have played their own important, but barely understood, role in redefining the geometry of urbanisation. For Brenner (1999), Keil and Ronneberger (1994) and Soja (2000), the enlarged scale of contemporary urbanisation, and with it the emergence and growth of post-suburban areas, bears only a partial resemblance to the Chicago school's radial and concentric geometric depictions of urbanisation. As Keil and Ronneberger suggest, 'instead of the radial-concentric concept of urban space, the notion of a nodal, fragmented pattern of relationships in a disparate urban fabric, with diversely dimensional cores and peripheries, seems to be taking hold ...' (1994: 139).

In this respect, the Ghent Urban Studies Team (1999) argue that the spatial reorganisation of urban areas is, if anything, more complex in the European setting with post-suburban areas having significant autonomous linkages elsewhere in wider city-regions. On the one hand, post-suburbia is usually defined in relation to the cores it surrounds but, on the other hand, it has lateral relationships with other suburban and post-suburban developments. Yet, unlike the classic models, post-suburbia rarely embodies a cohesive territory. Instead, post-suburbia displays tendencies towards internal fragmentation and can be the focal point of boundary transgressing institutions and processes. Indeed, as the sub-title of Garreau's (1991) book suggests, the most dynamic of post-suburban areas are at the constantly shifting frontier of contemporary urbanisation processes.

The term 'edge city' is in fact rather loose and conceivably many suburbs or post-suburban areas would conform to at least some, if not most, of the five criteria. Moreover, there is little that is explicit in Garreau's definitions or descriptions about the characteristic form of edge cities. Instead, and perhaps not surprisingly, this vacuum has been invested in the minds of scholars with the specifically American urban form of low-density developments at the confluence of major road connections. Whilst edge cities have a recognisable centre, it is also clear that the form and symbolic content of this centre presents a contrast to the focal points of established cities. As Garreau describes, Edge cities' 'characteristic monument is not a horse-mounted hero in the square, but an atrium shielding trees perpetually in leaf at the cores of our corporate headquarters, fitness centers, and shopping plazas' (1991: pre-introduction).

The term 'edge city' may not even be suitable to capture the nature of post-suburban development since, 'in defining edge cities primarily by their centers, those spaces in which commercial or retail activities occur, he [Garreau] . . . ignores the fundamentally decentered or multi-centered nature of these emerging regions' (Kling, *et al.*, 1995: xiv). Instead, Gottdiener and Kephart argue,

> urban life is now organized in metropolitan regions composed of polynucleated and functionally differentiated spaces that are no longer extensions of the traditional city. They are neither suburbs nor satellite cities; rather they are fully urbanized and independent spaces that are not dominated by any central city. (1995: 34)

In what has been depicted as a post-modern urban landscape of multiple and competing 'intentionalities' associated with a range of actors with only partially coincident interests (Dear, 1996), the 'modernist ideals and instruments of planners appear as relics such that ' "city centres" become almost an externality of fragmented urbanism; they are frequently grafted onto the landscape as a[n] . . . afterthought by developers and politicians concerned with identity and tradition' (Dear, 2003: 503). The case of Espoo, presented in Chapter 7, with its five separate centres including its rather artificial administrative centre, provides a vivid example of just this feature of post-suburbia.

Perhaps as a result, subsequent terminological innovations have attempted to capture this diffuse nature of urbanisation in which even some of the centres lack a sense of place. These nodes composed of the likes of retail and commercial developments and airport-industry complexes represent non-places (Augé, 1995). Taken together they constitute the 'geography of nowhere' (Kunstler, 1993). The geography of nowhere nevertheless shares a lineage with or, at the very least, a relationship to older processes of suburbanisation as Kunstler makes clear. 'The streetcar lines had promoted suburbs of a limited scope, a sort of corridor out of the city . . . The auto now promised to fill in the blanks between the streetcar corridors, and then to develop open space far beyond the city limits' (Kunstler, 1993: 89). The term 'tech-noburb' deployed by Fishman also highlights the difficulty of delim-iting today's sprawling post-suburban settlements. 'Compared even to traditional suburb, it at first appears impossible to comprehend. It has no clear boundaries; it includes discordant rural, urban and suburban elements' (Fishman, 1987: 203).

The term 'edgeless cities' has been used by Lang (2003) to describe edge cities without a sense of place that nevertheless account for an estimated two-thirds of non-downtown office space. Such edgeless cities 'are not even easy to locate because they are scattered in a way that is almost impossible to chart. Edgeless cities spread almost impercept-ibly throughout metropolitan areas, filling out central cities, occupying much of the space between more concentrated suburban districts and ringing metropolitan areas' (Lang, 2003: 1–2).

What of Europe? Does it have anything to compare to these new urban forms? There is a difficulty in identifying genuine points of comparison given what might be referred to as 'disparities' in both the geographic scale and the history of development at the urban edge. Nevertheless it could be argued that European metropolitan areas have also experienced some elements of urbanisation apparent in the US cities including the decentralisation of employment, the growth of car ownership and the growth of office and retail parks. As such 'the difference between the North American and European city seems to be one of proportions than of substance; in the USA... changes have been much more extreme and extensive' (Mazierska and Rascaroli, 2003: 18). There is a case for arguing then that this 'dimensional disparity' obscures at least some valid points of comparison between post-suburban forms in different settings. Nuissl and Rink (2005) have noted the heavy involvement of real estate companies and anonymous investment funds in the produc-tion of urban sprawl in eastern Germany and its partial similarities with Fordist-style residential suburbanisation in the US. Similarly, Bontje and Burdach argue that 'recent development tendencies in European metro-politan regions bear resemblance to Edge City development in several respects' but are ' "typically European" variations on the original Edge City model' (2005: 317).

One might add a 'temporal disparity' to the dimensional disparity noted above – that is, differences in the pace and timing at which such post-suburban settlements have emerged in different settings. As Jane Jacobs (1970: 48) noted some time ago, 'memory does not go far back enough to dissemble appearances of modern urban form and function'. Something of this sentiment is also captured by Whitehand and Carr (2001: 121) when they highlight how the historical inevit-ability of adaptation of the built environment and its forms 'makes our own time quite unremarkable' yet at the same time 'sequences of change in the urban landscape have been the subject of... different conceptualizations'. Dick and Rimmer (1998) argue that, seen in comparative perspective, cities embody a set of elements that are bundled and unbundled in different settings and that there have been periods when the pattern and processes of urbanisation in North American and East Asian cities have converged, most notably at present.

Dick and Rimmer (1998) are noticeably quiet in positioning contemporary patterns and processes of urbanisation in Europe with respect to this trend of convergence. Yet some of the same urban forms are the subject of Sieverts' use of the term 'Zwischenstadt' in the European setting but which he also sees emerging worldwide, especially in areas where traditional city forms have not taken firm hold. He argues that despite

> the massive differences in the forces behind urban development, the result in each case is the diffuse form of Zwischenstadt, which separates itself from the core city – if one still exists – and achieves a unique independence. These characteristics link the area of Greater Tokyo with the Ruhr area, São Paulo with BosWash... and Mexico city with Bombay. (Sieverts, 2003: 6)

The lineage of forms

Yet the geometry of progressively decentred forms of urban development was perhaps never as rigid as some of the literature implies. Although with a specific concern to analyse the rise of industrial suburbs, Walker and Lewis's broader historical purview enables them to identify this.

> The conventional explanation of industrial location in the city and suburbs has serious problems... First, nineteenth-century transport nodes were not as fixed or as nodal as is commonly asserted.... Second... transport access is often the *dependent* variable in the equation of industrial location. (Walker and Lewis, 2001: 6, original emphasis)

Moreover, whilst many suburbs have, until recently, remained economically dependent upon the central cities they surround, Lewis has noted the wider geographic orientation of early industrial suburbs arguing that their 'success... depended upon their relationship with other parts of the metropolitan area and with regional and national markets' (Lewis, 1999: 159) and that some of the earliest industrial suburbs had significant national and international connections.

Finally, just as the function of post-suburban areas has mutated over time, so too might one argue has the form of post-suburbia. Louis Wirth, associated with the monocentric and self-contained industrial metropolis of the Chicago school, was nevertheless able to observe not just the emergence of ex-urban developments but their mutation in form into the contiguous outline of the modern metropolis.

> The city and its hinterland represent two phases of the same mechanism. One of the latest phases of city growth is the development of satellite cities. These are generally industrial units growing up outside of the boundaries of the administrative city, which, however, are dependent upon the city proper after the city has inundated them, and thus lose their identity. (Wirth, 1925, cited in Dear, 2003: 502)

Placed in longer-term historical context, then, there is indeed a continuity in the geometry of post-suburban development.

> As cities have grown, larger upon larger suburban development has been added to the built-up area, leaving former outlying districts well inside the metropolis and often erasing historic patterns of expansion by dispersion in the process. After many years, it is easy to mistake the older edge cities and secondary nodes for part of a single 'central city'. (Walker and Lewis, 2001: 7)

A broader historical view such as that adopted by Walker and Lewis is likely to uncover such complex geometries of post-suburban development, which are likely to be closely related to mutations in the functions and symbolic content of suburbs.

The porosity of the urban edge

The porosity of urban forms and mixing of uses is a theme that can be found repeatedly over time and in different geographical contexts though it is one which has often been spoken of in exceptional terms. Agricultural uses were common place either within the tight confines of the pre-industrial city or just outside it (Sjoberg, 1960: 36). Writing in the early 1900s and from the perspective of the modern industrial city of the Chicago school, H.G. Wells predicted the intensification of this mixing of land uses in cities of the future.

> the city will diffuse itself until it has taken up considerable areas and many of the characteristics of what is now the country... The

old antithesis will indeed cease, the boundary will altogether disap-
pear . . . There will be horticulture and agriculture going on within the
'urban regions', and 'urbanity' without them . . . (H.G. Wells quoted
in Sieverts, 2003: vii–viii).

By the 1960s but at a broader geographical scale, Gottman was indeed
noting the mixing of land uses within megalapolitan form that had
emerged on the eastern seaboard of the US. The non-contiguous or
perforated form of major urbanised regions is something that, sporadic-
ally, has been returned to in often quite specific branches of the urban
geography literature. McGee (1991), for example, claimed a considerable degree of distinctiveness for such mixing of land uses within the
extended metropolitan areas of south-east Asia. These *desakota* developments mixed urban and rural uses, however, 'on the whole, these
zones are much more intensely utilised than the American megalopolis'
(McGee, 1991: 17). The porous nature of the constellation of separate but
functionally interlinked towns, cities and suburbs of the heavily urbanised south-east of England is something that, Allen *et al.* (1998) claim,
necessitates the term 'regionalisation'. Arguably, then, both urban and
rural elements are apparent more generally in today's modern sprawling
urban forms – what Sieverts terms 'Zwischenstadt' or 'cities without
cities' – in Europe and elsewhere.

The new found gravity of the urban edge?

Geographic scale is central to the manner in which the entrepreneurial strategies of post-suburban governments, politicians and other
non-government agents are played out. Such edge entrepreneurialism
centres not merely on material and discursive practices which establish
a sense of place, but since that sense of place is felt most acutely in
the context of broader metropolitan areas, it is played out in relation to
other geographic scales and senses of place. First, the pursuit of political, economic and social autonomy has been an enduring theme in the
material and discursive construction of North American (Teaford, 1997)
and European suburbs from the early 1900s to the present day.

Second, this search for absolute independence from central city or
metropolitan political and administrative arrangements has itself a
complex relationship with the varying degrees of relative independence that edge areas have displayed over time in economic terms (Hill
and Wolman, 1997; Savitch, 1995). Whilst some of the newer post-
suburban areas in North America display signs of increasing economic
dynamism independent from the central city and metropolitan areas

to which they are proximate, the more common pattern appears to be one of complex and selective interdependencies. For example, it is quite possible for post-suburban areas to remain largely dependent upon central city economic activity (either as a source of surplus labour or in terms of tertiary-sector divisions of labour manifest at the urban or urban system scale), while simultaneously exerting their own economic effects laterally upon other suburban and post-suburban areas.

Third, most recently, and perhaps as a consequence of the meshing of political aspirations with socio-economic conditions in some post-suburban areas, attempts to enlarge their 'spaces of engagement' (Cox, 1998) have begun to figure prominently in the entrepreneurial strategies of agents of post-suburban governance. In this way, important locally dependent institutions – notably municipal governments – are concerned to transcend their jurisdictions. It seems reasonable to suggest that such enlargement of spaces of engagement has a long, if uneven, history. So, for example, the economic basis of pre-industrial cities was secondary to and consequent upon the build-up and exercise of administrative and political power which was also the key determinant in the expansion of cities outwards into their non-urbanised hinterlands (Sjoberg, 1960: 68). Moreover, as Jacobs notes, the capitals of ancient city-states and empires were capitals precisely because 'they were large enough and strong enough to export their city governments, first to their hinterlands, beyond the home territory and then frequently further' (Jacobs, 1970: 143). The enlargement of the spaces of engagement of post-suburban municipalities may be more recent but is nevertheless part of such a longer-standing feature of urban government.

Instances of local institutions or coalitions of institutions expanding their spaces of engagement and extending their spheres of influence beyond their own jurisdictions and into the wider metropolitan space are apparent today. Indeed, Althubaity and Jonas (1998: 149) argue that suburban entrepreneurialism in the US may have rested crucially on the ability to access non-local public resources, specifically the 'ability of local government in suburban areas to harness state redevelopment powers . . . to lever inward investment'. As a result, post-suburban communities have become important centres of gravity within wider metropolitan spheres since, as Keil notes, 'the suburbanization of urban politics – whether in the form of urban secessionism or regional consolidation – has created a new political platform from which powerful political and economic actors . . . operate region-wide' (Keil, 2000: 758–781). In Europe the entrepreneurial activities of suburban governments seen in the likes of place marketing initiatives have generally been weaker

(Ward, 1998) but may also simply be a product of the sort of temporal disparity or time lag associated with a more gradual accretion of economic and political mass. The engagement of post-suburban governments with institutions at wider scales is something that we explore specifically in the cases presented in this book.

In sum what becomes apparent here is the fact that the very form of post-suburbia is constituted through a complex meshing of the actions of private- and public-sector agents within the development process – with some of these actors appealing to and successfully managing to mobilise non-local material and ideological resources. As Hise notes in relation to understanding the diffuse urban growth of Los Angeles, 'what is needed is an elastic analytical and interpretive framework that can expand to the region and contract to the district or neighborhood level and encompass points in between. Elasticity is important because as the community builders realized, no event or intervention takes place at only one scale' (Hise, 1997: 12).

2.6   Conclusion

One major purpose of this chapter has been to argue that there has been an overemphasis upon the *form* of post-suburban settlements which, as the outcome refracted through nationally specific institutional contexts, tends to militate in favour of exceptionalism in the study of the urban edge. This is seen most clearly in the fact that the term 'edge city', despite its loose definition, is now so firmly invested with a sense of the *form* of US urbanism that its use has obscured points of potential comparison between post-suburban developments in other settings.

A second purpose of this chapter then has been to redirect attention away from the appearances of form to the processes of growth and the functioning of post-suburban settlements. This leads on to a need to be sensitive to the range of different actors involved in the development process and their variable relations to economic structure (Ambrose, 1994; Healey and Barrett, 1990). The agency of local actors driving urban politics stressed within the US literature on urban regimes and growth machines needs then to be allied to a wider context, understood in more abstract theoretical terms of the variable structural position of different actors and their interests (Ward, 1996). The empirical material presented in Chapters 4–8 is partial in this respect. While we have been able to uncover something of the roles of businesses, land owners and property developers and residents in the development of post-suburban Europe, we have less to say, for example, about the role of financial investors.

Third, then, we have tried in this chapter to outline a set of themes that open up several avenues of analytical inquiry regarding post-suburbia in comparative perspective. This undoubtedly falls some way short of a cohesive theoretical framework linking structure and agency in the analysis of urban politics. Nevertheless, our themes draw attention to what abstract theories of the structuring of the state and interests can lend to an analysis of the origin and growth of post-suburbia alongside considerations of the agency of different actors in the urban development process more familiar within the analysis of urban politics (Molotch *et al.*, 2000; Ward, 1996). In the light of their longer lineage and variable national institutional settings, the specificities of post-suburban Europe have actually dictated the need for such a broader analysis of their form, function and processes of growth. In this respect, and in the light of our wish to dispel the current westward leanings within the urban studies literature, we have wanted to consider the relevance of elements of post-suburban function and form, not only from North America but from Europe and also East Asia.

Finally, in seeking such a broader analysis it should be apparent not only that elements of the seemingly rather exceptional edge cities of the US may be apparent within Europe (and elsewhere) but also that such a broader analysis also opens up new lines of inquiry regarding the origins and functioning of edge cities in the US setting.

3
In Search of a European
Post-Suburban Identity

> We didn't think at all about the American concept.... the fact
> that it was an American concept... we saw as an advantage
> because in a way, by saying "edge city", people would react to
> it – "oh, you're not an edge city".
>
> [Interview C9]

3.1 Introduction

Along with the likes of public–private partnership working, transnational networking is commonly cited as one means of coping with the
increasingly stringent fiscal climate in which European municipalities
have had to operate over the past two decades or so. The European
Commission's funding of trans-European networks has created new
sources of finance for policy development and implementation across
a wide range of spheres. The practice of networking itself arguably
offers important knowledge spillovers to participating local authorities
and important possibilities in the development of common European
identities – though significantly these broader benefits may be negated
by powerful forces of inter-locality competition for private and
public investment and by deeply ingrained national patterns of inter-
organisational working.

In this chapter we consider the emergence of one such European
network – the recently formed edge cities network. Following Dejan
Sudjic's (1993) likening of Croydon to Garreau's (1991) notion of a US-
style 'edge city', the term was adopted by members of the network.
However, although several of the member municipalities share *some* of
Garreau's five defining features of US-style edge cities, the use of the
term within this network was more a piece of opportunism. As the

opening quotation above indicates, from the perspective of officers in Croydon Council instrumental in establishing the European network, the American concept of an edge city was a foil to their own designs to raise the profile of the municipality.

We begin this chapter by considering the growth of trans-European local authority networking noting the issues regarding the formation of shared identities which transcend national territories in Europe as well as those surrounding the direct and indirect benefits and nature of networking activities. We then pass on to discuss the formation and practices of the edge cities network, drawing on original empirical material drawn from our five case-study post-suburban municipalities – Kifissia, Getafe, Noisy-le-Grand, Espoo and Croydon.

3.2 Trans-European local authority networking

Trans-European networking among local authorities dates back to the late 1950s but has developed most rapidly since the early 1990s. Perkmann estimates that there are now over 70 instances of such trans-European networks. Such trans-European networking includes the establishment of collaboration among contiguous sub-national authorities to form cross-border regions (CBRs) and what Perkmann (2003) terms 'interregional co-operation' among geographically non-contiguous authorities. The edge cities network, which is the subject of this chapter and from which our case studies of post-suburbia in this book are drawn, falls into the latter category.

Moreover, initial efforts at trans-European networking were made under the auspices of the Council of Europe which promoted the legal frameworks for collaboration within CBRs. Over the years this has given way to transnational networking stimulated and driven by new streams of finance made available by the Commission as part of its regional policies. As such, many CBRs, and presumably interregional co-operation arrangements such as the European edge cities network, have been seen to 'function as implementation agencies for this specific type of transnational regional policy' (Perkmann, 2003: 155).

The RECITE programme, launched by DGXVI in 1990, was created by siphoning off a small amount of funds under Article 10 of the European Regional Development Fund. The programme co-financed 37 networks and was expanded in RECITE II (1995–1999). A 1993 amendment to Article 10 secured a more substantial (1 per cent top-slice of annual European Regional Development Fund (ERDF) budget) and permanent funding stream for a series of trans-European networks (Rees, 1997: 391).

The various trans-European networks of local authorities across Europe can be categorised under several different headings. First there are local authority networks which are based upon sectoral co-operation. The first wave of RECITE funding encouraged co-operation between municipalities whose economies have been hit by restructuring in the same dominant sector. These networks include MILAN (the motor industry local authority network), EUROCERAM (ceramics) and DEMILITARISED (defence industry) and brought together local authorities with experience of coping with economic crisis. RECITE II (under which the edge cities network considered here has been funded) threw off this sectoral focus to offer a broader range of five themes under which projects were eligible for funding.[1]

A second category of networks are essentially based around lobbying activities. Local authorities quickly realised that lobbying for EU funding or to influence policy has more effect when done as a representative grouping rather than as an individual authority. Moreover, the European Commission has been favourably disposed towards local authorities, as a means of reducing the 'democratic deficit' held to be affecting the Union's legitimacy.

A third category of networks are concerned with issues of spatial planning and development. Many authorities are developing 'clusters' or macro-regional groupings of geographically neighbouring economies to facilitate co-operation in the spheres of place marketing, cultural and education facilities, infrastructure development and so on.

Networks centred on more prosaic concerns of enhancing service provision represent a fourth category. As service providers or enablers, local authorities can benefit from transnational exchange of experience in areas of common interest: this could involve the creation of jointly funded pilot projects, such as that dealing with traffic management (POLIS), the creation of databases which document approaches to problem-solving, or it may involve exchange of experience through seminars, conferences or study visits.

Finally, we can also note the role of big city organisations such as Eurocities or the Council of European Municipalities and Regions (CEMR) in helping to influence the EU policy agenda, feeding into the representation of urban authorities in the Committee of the Regions (rather than confining that body simply to regional authorities). The Eurocities network might represent a fifth category of municipal groupings that, in part, embody a place-specific synthesis of common interests across some or all of the other four spheres.

As the number of these trans-European networks has grown from the 1950s, there have been several significant changes in their characteristics. So that,

> It is symptomatic of European integration in the post-war period that the more legalistic approach favoured by the Council of Europe – proposing CBRs as formal politico-administrative entities – was later abandoned in favour of a more pragmatic and economically oriented approach within the context of EU regional policy. (Perkmann, 2003: 155)

Pursuing this theme, Le Galès has summarised the diversity of effects that transnational networks have had on urban politics and administration across European states.

> These transnational networks are privileged sites for obtaining information, exchanging experiences, ideas, and knowledge of various kinds . . . Individuals . . . very often mention the importance of these networks in . . . not only understanding the dynamics of policy-making at the EU-level, or how to obtain funds, but also making sense of new repertoires and norms, understanding the logic of uncertainty associated with some programmes, understanding the dynamics of coalition-building and the diversity of interests represented within the EU, and not least, understanding profoundly different institutional settings. (Le Galès, 2002: 107)

The impact of these networks is perhaps at its clearest and most direct in terms of the significant build-up of European-related human resources at municipal level (Perkmann, 2003: 157). Drawing on these observations, some further specific issues relating to these networking initiatives merit closer inspection, notably: the relationship of networking to common European-wide identities; the direct and indirect benefits from networking among local authorities; the barriers to generating such benefits from networking; and what the former imply in terms of the balance of co-operation and competition among local authorities in Europe. We consider each of these in the remainder of this section.

Networking and trans-European identity formation

Atkinson (cited in Chorianpoulos, 2003:688) suggests that Commission-funded local authority networks assist in the development of a 'common discursive context'. According to Friedmann (2000), the Eurocities

network, for instance, was not especially inspired by economic development objectives but was imbued with the aims of European integration
and social cohesion with significant interest in forging a specifically
transnational identity among participating cities. The emphasis in many
other transnational networks, however, has been strongly oriented
towards economic development objectives and arguably reflects a
broader but decisive shift in the nature of urban politics (Harvey, 1989;
Mayer, 1995). The creation of any 'European post-suburban identity'
is therefore likely to be rather incidental to a network within which,
given the main focus of the RECITE programme, economic development
objectives are important alongside those of social cohesion. Moreover,
the diversity of municipalities included in the network to some extent
militates against the forging of a common European post-suburban
identity.

Nevertheless, one Commission official concerned with administering
transnational networks was apparently able to comment that authorities have 'begun to discover how to co-operate between themselves in
areas where there may be a common problem to a number of different
regions . . .' (quoted in Rees, 1997: 402). Here, the degree of commonality
among authorities existing at an operational level may differ from that
at a strategic level, not least because the former may be tightly prescribed
by particular funding streams. The degree of common identification
with strategic and operational objectives of municipalities within the
Eurocities network, for instance, appears strong. These twin strategic and
operational goals are also present in the edge cities network considered
here. But as we shall see, identification of municipalities at a strategic
level, indeed over the very essence of what a European edge city is,
has tended to be quite weak. Moreover, whilst funding for networking
can bring benefits on an operational front it may also circumscribe the
identification with common strategic objectives.

The benefits of networking

There may be direct and indirect benefits or spillovers in that formal
funding of networking activities may also promote the exchange of
knowledge and practices in related projects or spheres. The transfer of
know-how among network partners is a specific objective of RECITE II
networks under which it is argued that

> Inter-regional co-operation offers a forum for working together
> in order to compare ideas, methods and practices. All involved
> can . . . learn lessons from their partners in other regions which will

enrich their own ways of working. In addition, exchanges and co-operation help to solve problems better and more swiftly. (CEC, 1996)

In this respect, the benefits expected are both direct and indirect and, *inter alia*, include: (1) Achievements which provide support for the exchange or transfer of know-how; (2) The development of a culture of co-operation; (3) Introduction of new methods or the improvement of existing methods; and (4) The introduction of a permanent structure to encourage the continuation of co-operation after financing has finished. Such transfer of knowledge is also considered a major benefit of inter-authority networking (Church and Reid, 1996; Rees, 1997). It is presumed to occur in relation to activities not funded by the Commission and to be predominantly from south to south or from north to south among network partners within the EU. Yet, the very narrowness of such projects and the short-term nature of funding involved can prevent longer-term broader indirect benefits being reaped.

Given the various types of inter-authority networks that exist, Commission funding is primarily for quite narrowly defined projects under which relatively modest levels of funding are available for tightly prescribed uses. In purely financial terms, there remain questions over the significance of resources available through the Commission's funding of trans-European local authority networks. Even in the prolonged period of increased fiscal stringency at sub-national tiers of administration witnessed since the 1970s in many developed nations, the relatively modest monies 'top-sliced' from the ERDF remain overshadowed by more significant funding available from respective national governments. Little surprise, then, that perhaps the key knowledge sought and transferred within networks concerns sources of funding and the means of accessing them. In this respect, trans-European networks provide vehicles for repeat funding from bodies like the Commission as a product of the exchange of expertise and best practice (Hebbert, 2000).

Consideration of the direct and indirect benefits of networking also raises broader questions regarding the nature of the inter-authority relations promoted through networking and their meshing with other essentially local initiatives pursued by individual municipalities.

Barriers to networking

There are broad political and administrative differences which create cleavages between northern and southern European municipalities in the network in terms of their participatory capacities. The southern member states typically have smaller municipalities with fewer competencies. Such differences have been seen by Perkmann to have had an effect specifically in the operation of transnational networks.

> Northern European local government has higher margins of discretion and a broader set of responsibilities, backed up by locally raised resources, compared to Southern European local governments' lower levels of discretion and responsibilities. It appears that these institutional differences are partly responsible for the ability of local actors to group together and form strategic coalitions across borders. (Perkmann, 2003: 165)

Moreover, the highly variable nature of national politics and the activities of political parties, in particular, present a different context to European urban politics (Harding, 1997; Herrschell and Newman, 2003). There are significant, broadly north–south, differences in the importance of elected political leaders – in the form of local mayors – when compared to bureaucrats – in the form of local authority chief executive officers. As Chorianopoulos summarises,

> The underdeveloped local administrative capacity and the limited presence of private and voluntary sector interest groups justify the concentration of authority in the position of the Mayor. Personified administration, in turn is evaluated through the mediating ability of the mayor to translate local authority into national policy influence, mainly through the party mechanism. (Chorianpoulos, 2003: 675)

Moreover, there are important geographical variations in the style of leading municipal bureaucrats in Western local government and within Europe (Klausen and Magnier, 1998b). Beyond these, differences in culture and language have been found to hamper co-operation within inter-authority networks funded by the EU (Rees, 1997: 400).

Networking and the balance of co-operative and competitive relations

An important point of analytical interest surrounds whether the fact that 'the Europe of strong states and "dependent" cities' (Harding, 1997: 296) also exerts an effect on identity formation and benefits produced from transnational networks. As Rees (1997) notes, the process of transnational inter-authority co-operation is an uneven one with differing desires and abilities to co-operate and differing capacities to act independent of higher administrative tiers. A key empirical question centres on the way in which transnational networks mesh with the changing urban institutional and political scene locally – being part cause, part effect of such changes. The picture of urban politics across Europe is summarised by Harding who suggests that 'while . . . the balance has changed somewhat in recent years as a result of privatization, deregulation and state restructuring, it remains the case that coalitions in European cities often tend to be public–public rather than (or as a precursor to) public–private . . .' (Harding, 1997: 300). The Commission's own reviews of transnational networking have argued the need for greater public–private partnership (Rees, 1997: 397).

Somewhat of a revival of competition between towns and cities has been coupled with processes of European integration (Cheshire, 1999). It seems fair to assume that trans-European local authority networks will reflect, and even be an important vehicle for, such competitive processes. Thus Le Galès argues that trans-European networks 'now reflect a Europe within which political and economic entrepreneurs try to promote cities in Europe in order to gain in terms of both power and economic development' (Le Galès, 2002: 108).

Here, competition among networks of regions and municipalities does not simply focus on the pursuit of the likes of mobile investment, but plays itself out over broader economic and political objectives as in the Transmanche experience (Church and Reid, 1996: 1313). In their discussion of the potential of networks to reduce competition between cities, Leitner and Sheppard (1999) point to the danger of co-operation remaining 'instrumentalized to the logic of competitiveness', in other words merely perpetuating a form of competition through collaboration. In this scenario, participating councils may be using the enhanced knowledge and policy best practice to distance themselves from less 'networked' neighbours, perpetuating the idea of 'warring brothers' developed by Peck and Tickell (1994) – so much in common, yet by that very fact doomed to quarrel and struggle with each other. Thus, an important aspect of inter-authority competition centres on the marketing activities of municipalities, with UK authorities, in particular, seemingly disposed towards using networks to raise their international profile (Church and Reid, 1996: 1310).

3.3   The European edge cities network

The present network of nine partner municipalities (Figure 3.1) was initiated by staff from Croydon Borough Council in 1995 after a Europe-wide

Figure 3.1   Map of Europe with edge city network members

meeting of council officers. It was established informally at first with
an original eight partners self-funding bi-annual meetings with a view
to gaining European Commission funding. To date, subsets of the participating municipalities have been successful in gaining funding under
the Commission's REACTE, RECITE II and Culture 2000 programmes.
Members of the network responded to a call for applications for funding
under RECITE II originally made in 1996 (CEC, 1996) but did not obtain
funding until 1998. During this period one member (one of our case-
study municipalities – Noisy-le-Grand) effectively withdrew from active

participation in the network. In 2000 a Danish municipality (Ballerup) was added to the network as a result of previous contacts with the existing Swedish member Nacka.

Through a regular cycle of meetings, representatives of the participating municipalities were able to identify a set of common issues over which co-operation might be established. The commonalities among these post-suburban municipalities centre on their inextricable relationship with the capital cities to which they are at least partially tied. The aim was clear:

> whilst these Edge towns and cities may have different patterns of development, what they all have in common is their proximity to the capital and the consequent need to develop a strategy which is based on a synergy with the capital city, but which also establishes their *separate identity*... (Edge Cities Network 1996: 2, emphasis added)

The RECITE II funding was eventually secured in 1998 specifically for a 3-year project centred on support for the internationalisation of Small- and Medium-sized Enterprises (SMEs) (covering seven of the edge city partner municipalities) and social exclusion (covering just a subset of Croydon, Nacka and Fingal County).[2] RECITE funds accounted for a total of €2,391,566 of the total budget of €3,882,594 agreed for the project (Croydon Council, 2001a).[3] An additional year was added to the lifetime of the RECITE II project due to slow take-up of available funds.

We now pass on to describe some of the ways in which these metropolitan municipalities have been able to use European partnership working to develop joint projects. In doing so, we concentrate on empirical material gained from a study of our five case-study members of the edge cities network – Kifissia, Getafe, Noisy-le-Grand, Espoo and Croydon as well as extra information from North Down Borough Council in northern Ireland. Issues of urban governance were considered in an initial ESRC-funded study of three of these municipalities. Croydon and Getafe were selected as two of the more active participants in the network and with a view to gaining some insight into possible contrasts and knowledge transfers between Europe's north and south. Noisy-le-Grand was chosen partly in order to understand some of the limitations of trans-European networks. Although still a (rather inactive) member of the edge cities network, the municipality withdrew from efforts to gain funding under RECITE. The municipalities of Kifissia and Espoo were added in a subsequent British Academy–funded study examining business representation in urban politics, with a view to augmenting north–south comparisons within the European setting.

In what follows, we begin by discussing the issue of trans-border identity formation. We then go on to consider the balance of co-operative and competitive practices evident in the edge city network from the perspective of each of our three case-study municipalities.

3.4   Towards a European post-suburban identity

The partner municipalities vary greatly. Our five case-study municipalities alone differ considerably in terms of their socio-economic and demographic complexion (Table 3.1). With a population of 335,000 or so and a significant commercial centre, Croydon is the largest of the municipalities included in the network. It also has a much longer history of urban metamorphosis than any of the other municipalities within the network. In purely superficial terms of the indicators reported in Table 3.1, Espoo might appear to come close to Croydon; however, the distinguishing feature is its extremely recent growth in parallel with the very recent experience of urbanisation in Finland. Although planned as the focal point of a new town growth pole, and despite seeing similarities between itself and Croydon, Noisy-le-Grand is only a fraction of the size (population 60,000) and, as we shall see in Chapter 6, its local institutions cannot draw on any clear sense of place identity and have none of the autonomy or self-confidence of those in Espoo or Croydon. The two south European municipalities – Getafe and Kifissia – are also

Table 3.1   Summary details of five case-study post-suburban municipalities

	Land area (km^2)	Population	Employment	% Tertiary employment
Croydon	91	335,000[c]	156,740[c]	82.2[c]
Espoo	528	221,600[a]	110,630[d]	82.7[d]
Noisy-le-Grand	13	60,000[b]	25,400[b]	87.0[b]
Getafe	78.8	150,432[c]	32,780[d]	57.0[d]
Kifissia	99.5	43,929[c]	18,897[c]	74.4[c]

Note:
[a] 2003
[b] 2002
[c] 2001
[d] 2000.
Sources: City of Espoo (2003); Ayuntamiento de Getafe (2002); Croydon Council (2005); Direction du Développement Économique et de l'Emploi (DDEE) (2003c); Hellenic Office for National Statistics (2004).

small but completely unalike. Getafe is different in having a very small commercial centre but being both a sizeable dormitory suburb and an industrial centre within the Madrid metropolitan area. Kifissia's original function as a resort has mutated first to an affluent dormitory suburb and most recently into a more mixed function settlement that includes also a select retail and office commercial centre and some manufacturing.

On the face of it such diversity would appear to militate against both the forging of a distinct pan-European post-suburban identity and benefits from networking among the member municipalities. Although at first glance, and as the members were themselves able to identify, there are some superficial similarities among the municipalities in the network, some limits to the creation of a common identity among European edge cities have also become apparent. As one interviewee from Croydon Council suggested, 'to say that there is a unique European concept of an edge city – there isn't – because we are all different, but we all found that we had enough in common to make it work' [Interview C9].

The lack of similarities with other members in the network proved enough of a problem for one of the original partners to eventually withdraw from the network. As an officer at Noisy-le-Grand identified,

> We didn't have enough in common. They were too different apart from the notion that they were 'edge cities'. We may have some things in common with Croydon, but not with the towns on the edge of Belfast, Dublin or Athens, for example. [Interview N1]

Yet diversity can itself confer opportunities for trans-European co-operation and policy development. Staff participating in the network in Getafe, for example, saw the diversity among partners as presenting opportunities for more numerous but focused partnerships [Interview G8].

Our interviews reveal a number of issues with which there is a degree of common identification among post-suburban municipalities in the network. The first of these was, as we saw above, made explicit at the outset in the formation of the edge cities network. It relates to a theme familiar in the history of suburban developments.

> There are certainly concerns about loss of identity.... They realise they are part of a wider city-region but they want to develop their own local identity and sustain it.... They want to create more local jobs and employment and a sense of community locally.... We just don't want to be part of an amorphous urban sprawl. [Interview ND1]

As we will see below, this in turn leads to the activities of the edge cities network meshing with other trans-European and local activities to become part of competitive outlook at least among some of the members.

A second issue around which there is some common identification among the members of the edge cities network concerns the existence of marginalised and young populations in many of these municipalities.

> Maybe one of the common features is that, because they are at the edge, it is about people flow. That the deprived or the asylum seekers get pushed to the edge because generally the central cities want to get rid of them, or they are coming in from somewhere else. So the airport might be located on the edge. So they become a centre of the migration. [Interview C18]

Again, this view makes a link between an old, even ancient, aspect of suburban development – that less desirable elements of population and urban activities are excluded from the city proper on the one hand – and comparatively recent developments such as airports that are associated with thoroughly modern experience of the non-place urban realm.

A third aspect common to network members concerns what might be termed an 'investment deficit' especially in relation to their capital city neighbours.

> I think there is an identity issue about the edge and the centre. I think there's a huge investment issue. . . . as edge cities, one of the big issues is the lack of political focus on where Government funding and Government investment is not in their areas; it's still in the centre. That doesn't mean that they haven't counteracted that because most of them have . . . Most of them have driven very strong public–private partnerships and that's the way they've dealt with it. [Interview C17]

What we see here are perceived limitations with the transport and communications infrastructure underpinning the connections of these post-suburban municipalities with their capital city neighbours and indeed urban centres further afield.

Taken together, these sorts of issues commonly felt by municipalities in the edge cities network reflect a broader belief in their 'invisibility' to central government and its major expenditure on schemes to address problems of unemployment, social exclusion and infrastructure improvements. As such, there is some evidence of an 'imagined community' among the network municipalities.

Irrespective of any such common identity, the vitality of the network must have rested on a commitment from the remaining members to working in partnership – especially since funding was not forthcoming for the first 3 years of the network's life. RECITE II guidelines resulted in a more circumscribed choice of project foci than would otherwise have been the case, and the whole process of securing funding put an enormous strain on the network. As one interviewee described,

> That was the problem with RECITE. It became very overwhelming for them... they had constant, constant justification and re-negotiation with the Commission. And people got very demoralised which was not to do with them or their own inefficiencies... And that in itself united them in the first place, but actually became quite disintegrating at one time because people started to feel quite negative. And that... left people feeling that there wasn't anything holding them together in the end. You know, what was the edge city stuff? [Interview C17]

However, this very process and the need to settle on a limited choice of projects generated a partnership of enormous strength among the various members. As another interviewee commented,

> We kind of built it bottom upwards so it was a bit messy. And the two areas were social exclusion and SMEs where people felt most comfortable with as a project... Ours was genuinely a collaborative effort where everyone kind of chipped in from the bottom. The advantage was that the partnership was extraordinarily strong... The first draft [bid for funding under RECITE II] I think was very strong on partnership but very weak on concept. And the European Commission bought into it because I think it saw the strength of the partnership. [Interview C9]

Here then the sorts of co-operative practices giving strength to the network were gained to some extent at the price of policy coherence.[4] One element of the RECITE II project centres on the partnering of SMEs from municipalities in the network with a view to supporting technology transfer, transnational business opportunities and the like. However, one interviewee suggested that important differences in the

sectoral-profile of business stocks across the participating edge municipalities have meant that suitable partner firms have been hard to identify
and the quality of match between those that have been identified is not
all that it might be. As a result, the lifetime of the RECITE II project was
recently extended by a year into 2002 due to some of the municipalities
being unable to take up funding to the levels originally envisaged.[5]

The European funding for the edge cities network is related to specific
projects. As the network has matured, this has begun to have implications for the wider benefits or spillovers of networking. Thus after
securing RECITE II funding

> there is a strong realisation that the network itself is not funded.
> RECITE funds RECITE which is only seven of the partners and for very,
> very specific activities. There was a fond illusion . . . that the funding
> we got through RECITE for transnational activities and transnational
> co-ordination and transnational meetings would in some way support
> the wider activity of the network. But in fact of course it isn't true.
> The Commission has placed a very tight framework around what that
> money can be spent on . . . And so a lot of the other activities fell off
> the end. [Interview C17]

From 2000, the network has funded its bi-annual cycle of meetings
from an annual membership fee. A subset of the network partners have
been involved in a small 1-year Commission-funded project. However,
after the 4-year project that was RECITE, it was suggested that partners
were now reflecting upon what sort of network they wanted to become.

The story of the formation of the edge cities network appears, as in
other examples of EU-funded networking, to confirm a level of commitment to networking even in the absence of such funding. The EUREGIO
Euro-region covering Germany and the Netherlands, for example, was
the subject of local authority networking dating back as far as 1958 and
continuing for some considerable time before funding became available (Van der Veen, 1993). The commitment forged prior to funding in
the case of the edge cities network is admittedly not of this order but
perhaps offers something of a contrast to other networks where longer-
term commitment to inter-authority partnerships seems to have been a
product of, rather than a prelude to, funding (Church and Reid, 1996;
Hebbert, 2000).

Overall, the picture that emerges here is that of a weak sense of
common European post-suburban identity having been forged from very
strong commitment to partnership working among a number of member

municipalities. Unsurprisingly, no explicit European definition of the term 'edge city' around which a strong sense of pan-European identity might have formed appears to have been forthcoming. However, there is a sense in which even such weak transnational forms of identification within external networks can be important in conferring identity upon individual urban areas. As Dematteis notes, 'an image begins to form of the city as a "node" of global networks, where local identity and the urban territory, as a stratified deposit of natural and cultural assets, no longer have value for what they are but for what they become in the process of valorisation' (Dematteis, 2000: 63). Arguably, their insertion into transnational municipal networks is all the more vital in the case of post-suburban areas for whom, it is clear from our discussion above, there are distinct problems of forging an identity separate from their respective capital cities.

3.5 Benefits of networking

There appeared to be examples of the indirect benefits that funding under the likes of the RECITE II programme was intended to stimulate in the edge cities network. These benefits appeared to be stressed, in particular, by the Southern European network members – Getafe and Kifissia – who spoke of an improved understanding of the cultural, political and administrative diversity of members (Gyro Consulting, 2002: 59). Thus for the edge cities network representatives in Kifissia 'actually it is communication and collaboration . . . at different levels. For us, that we work on such kinds of projects, definitely it's a benefit because we are learning different mentalities and ways of organisation' [Interview K8].

Perhaps the best example of an edge cities network member municipality acting as a source of best practices and exchanges of knowledge is presented by Getafe. Contrary to most assumptions regarding the direction of transfer of knowledge and best practice, Getafe itself was an important focal point for the rest of the network. As we describe in greater detail in Chapter 5, the municipality of Getafe could be taken as an example of 'edge entrepreneurialism', not least because of the political pragmatism of left-wing Mayor Pedro Castro. Getafe's involvement in the network also illustrates a certain pragmatism which approximates to the networking ideal of co-operation and genuine interest in exchange of best practice. Interestingly, Getafe's officers came to the network with considerable expertise (under objective two funding) with EU-funded activities centred on training for SMEs – experience of direct relevance to one of the European edge city network projects funded

under RECITE II. Despite Getafe being outside of the funding arrangements for edge cities under RECITE II, the network has nevertheless been
able to draw upon the municipality's established expertise. Moreover,
Getafe's acknowledged lead in promotional materials has seen it dissem-
inate these among the members of the network. As we saw earlier, a
major objective of networking funded under RECITE and a major benefit
perceived to flow from networking in general is the transfer of know-how
with such transfers occurring between southern regions and municipalities or from north to south. However, what Getafe's participation in the
edge city network appears to demonstrate is a significant south–north
exchange of best practice.

Echoing Hebbert's (2000) discussion of Transpennine local authority
networking, one of the themes to come out of our research is the role of
networking in conferring spillovers upon partner local authorities in the
form of additional knowledge and experience relating to the accessing
of further sources of funding, in this case from institutions like the
Commission. This was highlighted most clearly by council staff at Getafe
who saw the network as a base from which to involve themselves more
widely in EU-funded projects.

> The network is a platform for us to involve ourselves in projects financed
> by the European Union ... the possibility this network has is that
> you don't have to go looking for transnational members outside it. If
> you want to do a project you already have the partners. [Interview G8]

Getafe joined the network too late to become involved in RECITE
II-funded projects, yet the edge cities network was the source of
a narrower group of partner municipalities (including Getafe) who
obtained funding related to environmental sustainability under the
Commission's REACTE programme. Staff involved in the edge cities
network at Getafe also identified their own potential role in enhancing
the knowledge resources and geographical scope of the network when
suggesting their role as a conduit for knowledge transfer to and from
the Latin American setting [Interview G8].

As the network has matured, there is evidence that general knowledge
of each others environs and projects and the like appeared to inform
exchange visits between the municipality staff outside the RECITE
II-funded network as a direct result of these activities. The contacts and
relationships forged through this network have stimulated additional
exchanges of knowledge and practices in spheres outside those funded
through specific Commission schemes or those considered the thematic

priorities of members of the edge cities network (Gyro Consulting, 2002: 28). As one representative from the network explained, 'we've done a lot on the back of those relationships as well. I mean people have done a lot of visits and shadowing of expertise and those sorts of things that come out of us having a longstanding commitment' [Interview C18].

Echoing the same theme, another interviewee noted how these wider benefits themselves helped to sustain the network during times when project funding was not forthcoming.

> Outside of the scope of RECITE, there has been quite a lot of good practice unearthed elsewhere in the edge cities network. Which is the sustainable bit of the links maybe – that we are not totally dependent on the RECITE funding. [Interview ND1]

Nevertheless funding remains a key concern though one felt unevenly across members of the network with something of an imperfect north–south European divide. One representative from the network indicated that he thought that his municipality would not have continued to be a member without the RECITE II funding. He went on to highlight the state of play in 2002 at which point members had just failed in one bid and were awaiting news of another bid for funding.

> The fuel that keeps it together is *a* European project. It doesn't necessarily have to have all partners in it. The next big challenge is to find another European project that will get enough critical mass from the partners into it to help sustain the edge cities network. . . . Some of the partners, perhaps like Getafe and Kifissia, have more of a philosophical view of membership of the edge cities network – that it helps make Europe a smaller place and through interaction it helps people to understand each other and help prevent conflict. . . . Kifissia, Ballerup and Getafe and perhaps Loures seem to be happy enough on that level. But in the other places like Croydon, and Fingal and ourselves and probably Nacka and Espoo, their politicians are looking for some tangible evidence of what the benefits have been. [Interview ND1]

3.6 Barriers to collaboration and the transfer of knowledge

The edge cities network is not funded by the Commission as a network *per se*. This as we saw above creates its own problems in terms of the longer-term future for the network. In this context, the perpetuation and growth of the network is something that involves a delicate balance: a balance between diluting the degree of identification with the aims of the network, on the one hand, and the addition of new members and with them new ideas and sources of potential funding, on the other hand. As one interviewee highlighted,

> The sense is that the network can only sustain one new member at any one time. Partly acknowledging that we have already lost Noisy-le-Grand and while we haven't lost any other partners that may happen as a consequence of changes of policy or interest. And the network needs to take advantage of the possibilities of the accession countries as well. I mean in terms of funding opportunities but also in terms of just broadening our base really. [Interview C18]

Network members have therefore invested considerable energy in looking for new sources of funding. In this they have, as we have seen, been successful but such funding is for discrete projects which are specific to particular themes and covering subsets of the members. This in turn creates problems for the forging of a common identity and cohesiveness among members in the longer term. Problems are manifest irrespective of the scale and longevity of funded projects as one interviewee observed.

> The thing is that when you have a long programme it is much better for us because we need time to do things... just to start doing things... That's the problem with the small programmes that until we have all the things that we need to start it, it is almost the middle of the project so we have to run after that. [Interview K6]

Here the short-term project being referred to is one funded under the Commission's Culture 2000 initiative. But projects like that funded under RECITE II – *the* major project that has sustained most of the members for a large part of life of this network to date – generate their own problems too. The task of managing and co-ordinating such projects is complex due to the size and diversity of the network itself. So, for example,

> what we found out from big programmes like RECITE II is that when you have a lot of people, a lot of countries involved, it is very difficult to coordinate. Especially when the come from very different countries. You know it's very difficult to bring people together. [Interview K6]

Following on from this are related issues of administrative burdens. In the lifetime of the RECITE programme alone there have been increased administrative burdens as wider concerns with the accountability of EU institutions have filtered through to individual funding streams such as RECITE II.

> Another major problem we have had is... the increasing bureau-cracy of the administration with it at the expense of the [RECITE II] project. The more and more that all the partners spend reporting and monitoring and auditing and verifying things the less time and money it leaves to actually do things in terms of productive contacts with the SMEs and trying to achieve the objectives of the project. [Interview ND1]

There was some suggestion from our research interviews that such factors had a significant bearing on the workings of the edge cities network – including the effective withdrawal of one of the original members. Taken together, however, there is no simple north–south European pattern regarding how these cultural and language factors impinge on member municipalities.

The role of and affairs surrounding elected Mayors as political as opposed to executive officers within council matters is also very different from those in northern Europe. The network representative described the parallels that exist between local government in Greece and Portugal when trying to contact representatives from the Municipality of Loures.

> The other problem I have experienced is that all local authorities do not function in the same way.... I think this is the case for Portugal also. For example, in the elections... we could not reach them for two or three months. But we knew as Greeks that it was a problem that they wanted to see what the outcome was of the elections. [Interview K8]

To an extent these difficulties were recognised among those working in northern European municipalities as an interview with an officer from Croydon Council illustrated, 'the other thing is political change... either political change or political electioneering. Because cities retrench into this "we can't think about the world because we are thinking about our local politics"' [Interview C18]. The same interviewee went on to suggest that some partners were affected more than others by such political elections and that there was a feeling that such difficulties were not always recognised centrally within the Commission.

Variations in the institutional context of local government can also have broader impacts on cultures of working. Perhaps equally important as the above factors, then, was the perception of the remaining members in the network that the effective withdrawal of Noisy-le-Grand from the network was connected to the municipality's lack of commitment to partnership-style working. Certainly it appears that the concept of public–private partnerships has yet to take root in the French setting in quite the same way as, for example, in Britain. As one interviewee suggested,

> the other thing they had great difficulty with was this idea that you worked in partnership. Our experience with the French is that, on the whole, they don't understand public–private partnership. So for example, when we asked everyone to put together a list of all their partnerships, they didn't have any. [Interview C9]

However, seen from the French perspective of stronger public–public partnership, working with more definitive and substantial lines of public-sector funding may have prompted some frustration with the protracted period over which the remaining members struggled to obtain a relatively modest amount of funding. Central state involvement and direction in lower administrative tiers remain strong in France. Attempts to invigorate the regional tier of government, for example, appear largely to have failed precisely because of continued dominance of the central state within the likes of joint planning exercises (Newman, 2000). There is a sense here in which the typically grander style of French public-sector planning and administration is reflected in Noisy-le-Grand's flirtation with the network. Thus from the perspective of staff involved from Noisy-le-Grand 'the whole thing was amateurish and oriented towards securing finance for their own small projects' [Interview N1].

Moreover, staff at Noisy-le-Grand's newly formed economic development department clearly expressed their difficulty in participating fully with the activities of the network at a time when the municipality faced significant social, economic and infrastructure-related problems.

> RECITE wasn't ambitious enough for us, and the benefits insufficient compared to the investment necessary in terms of effort . . . So we

didn't really have the time to participate in RECITE, as the town needed a lot of work on it to rescue it. [Interview N1]

Beyond this lie more prosaic reasons for Noisy-le-Grand's disengagement from the network. The main strategies for economic development in Noisy-le-Grand are contained in a series of interlocking projects and plans, elaborated at various levels of governance as we describe in greater depth in Chapter 6. These plans and projects ensure the town will benefit from extra development funds to pay for projects within the local and regional plans as well as others, and reinforce the articulation between state priorities and local and regional development strategies. In this way, national priorities are cascaded down from the national to the regional and local levels. The uncertain status of the network and its prolonged search for modest RECITE funding can be compared with the more complicated but definitive contractual planning and funding arrangements. The gains from involvement in inter-authority networking vary according to the differing expectations of members (Rees, 1997). Clearly the case of Noisy-le-Grand reveals a series of specific national and local circumstances which contributed to expectations divergent to those of the remaining members.

Language barriers appear to operate on a broad north–south basis because of the predominant use of English as the working language of the network. Interviewees in both northern and southern European municipalities participating in the network identified difficulties experienced by southern members of using and translating to and from English. Yet there are important examples of language presenting barriers to effective communication and collaboration across northern European countries. There is more than a suggestion that this was a significant, albeit a specific, barrier between France and the other northern European member municipalities that contributed to the withdrawal of one member of the network. Instances of literal translation leading to misunderstandings among northern European members were noted [Interview C18]. Moreover, effective collaboration in trans-European networks is dependent not just on use of language but on communication defined more broadly. Here too, important cleavages among northern and southern European municipalities taken as groups can be apparent. So, for example, as a representative from Kifissia noted,

The Nordic countries speak very good English but they are not good in communication. You can send emails that are never replied to.

They are very limited in what they are going to say or what they are going to explain.... They have a very different mentality even from the English and Irish. I think they are the most difficult people to communicate with for us. [Interview K6]

3.7   Balance of co-operation and competition

As the municipality which took the lead in forming the occasionally Commission-funded network of European edge cities, Croydon Council's championing of the European network highlights the intrinsic ambiguity of the term 'edge city'. It also stands for a broader bias towards the competitive value of network membership among northern European network member municipalities when compared to a bias towards the collaborative value of network membership among southern member municipalities.

From the specific point of view of Croydon Council the interest in using the label edge city has its origins in longer and more firmly locally held beliefs in the borough's being a city in its own right. Croydon first bid for city status in the early 1900s and, according to its latest bid (Croydon Council, 1999), is the largest town in western Europe without city status. Here it seems post-suburban areas such as Croydon have made the same heroic appeal to history and municipal autonomy as their larger capital city neighbours within the Eurocities network (Friedmann, 2000: 127). As such, Croydon's self-promotional use of the term also bears little resemblance to the North American idea of an edge city.

There's this Croydon as a city... This kind of European city kind of concept that Croydon has. It wants to punch above its weight. It wants to be something it's not... The interesting thing about edge city is not the edge, it's the city. [Interview C9]

In this respect then Croydon's opportunistic self-styling as an edge city sets it apart somewhat from the other less populous and economically weaker members of the network. There is a sense in which the edge city network has been used by Croydon as an adjunct to efforts to market itself in wider terms, notably in the case of its recent bid for city status (Croydon Council, 1999; Meikle and Atkinson, 1997). A view that stresses the value of trans-European networking primarily in economic terms underlies these aspirations in Croydon council, and presumably in several of the larger municipalities within the network. As one interviewee explained, 'we have shied away very much from the twinning idea. We are not about civic community . . . I think it is very much an economic connectivity in the broadest sense' [Interview C18].

The edge city network is one among a number of partnerships which the council is proliferating. As staff at the council commented: 'increasingly we work as if we are "Croydon Plc" and we are building a partnership infrastructure which drives it . . . ' [Interview C9]. Here we see some confirmation of Church and Reid's (1996) findings that the co-operative practices of network participation can also be allied to competitive practices of place marketing. Moreover, it appears that networking rather than promoting significant changes to pre-existing local institutional structures or practices meshes with them (Rees, 1997).

A further implication here is that this partnership style of working is bringing with it an increasing emphasis upon the private side of the public–private equation. What the Croydon case appears to highlight quite strongly is a 'post-suburban entrepreneurialism'. Although at the time of writing New Labour controlled, Croydon Council's strong orientation towards the private sector dates back a long way under what has traditionally been a Conservative-controlled council (Saunders, 1983). This private-sector emphasis appears, to an extent, in the workings of other edge city partner authorities. However, the centrality of public–private partnership within local politics places Croydon a bit apart from other members of the edge cities network and stands in marked contrast to the continuing public-sector ethos apparent within the lapsed French edge city partner municipality. So, one recent study of regime politics in London boroughs, for example, concluded that 'Only Croydon begins to approximate a U.S. style of urban regime built around local economic development issues, bipartisanship and close public- and private-sector relations and partnerships' (Dowding *et al.*, 1999: 519).

Nevertheless, this interest in a post-suburban municipal identity as cities distinct from those of their larger capital city neighbours is something felt more widely. So much so in fact that for the more aggressive of network member municipalities the term 'edge city' carries unwanted connotations. As one interviewee in Espoo objected,

> I don't think edge city is a good word to use. This is Espoo. It may be near to Helsinki . . . but edge city . . . nothing. Some of those edge cities are like suburbs. And this is a city of its own and it is growing rapidly and it has its own economy. [Interview E8]

It is here that we see how the desire for municipal independence enshrined in the network's collective definition of a European edge city can easily spill over into localised and indeed international competitive aspirations of municipalities. Indeed, the continued commitment of the City of Espoo to the edge cities network has been called into question by its recent membership of the Eurocities network.

3.8   Conclusion

In this chapter we have presented some detail on the inter-authority relations in a recently formed trans-European network of post-suburban municipalities. The activities of this unique network highlight the often overlooked problems and possibilities facing municipalities at the edge of more illustrious capital city neighbours. Whilst the post-suburban municipalities that make up the network are indeed diverse in size and socio-economic complexion, the existence of some important commonalities and a strong commitment to partnership working have seen these municipalities forge something of a shared identity.

As Church and Reid have surmised, the sorts of cross-border co-operation found in the likes of the edge cities network can be viewed 'as a response to the complexity, fragmentation and privatisation of the local state' (Church and Reid, 1996: 1314). However, over much can be made of this given the relatively modest levels of funding involved and the limited and circumscribed nature of that funding. Moreover, the evidence presented here suggests that the direct and indirect benefits of networking are not all that they might be.

Our study of members of the edge cities network reveals some of the ways in which transnational networking does or does not mesh with changing local administrative and political contexts displaying varying degrees of complexity, fragmentation and privatisation. Whilst local and transnational coalition building can embody significant instances of co-operation among members, this study of the European edge cities network appears to suggest that the balance of competitive and co-operative practices is in favour of the former. The common identity and partnership working within the network did not appear to preclude the use of these network relations in autonomous actions on the part of individual local authorities in their respective metropolitan and institutional settings. Indeed, our study of this particular edge city network provided a window onto the sorts of post-suburban entrepreneurial political coalitions which remain understudied.

Finally, the nature and solidity of urban political coalitions across Europe is highly uneven (Harding, 1997). Thus, to an extent, this chapter also revealed, as one might expect, that patterns of inter-authority working bear the imprint of the long urban history and distinctive administrative and political traditions of nation-states in Europe. The most striking example of this within the edge cities network considered here was the lapsed French member municipality, Noisy-le-Grand. Here transnational networking did not sit at all comfortably with a complex and relatively non-privatised local administrative system.

4
Kifissia: Playground of the Athenians?

And as imagination bodies forth
The forms of things unknown, the poet's pen
Turns them to shapes, and gives to airy nothing
A local habitation and a name.

William Shakespeare,
A Midsummer Night's Dream

4.1 Introduction

In 'How Eden lost it's garden', Mike Davis (1996) recounts the destruction of southern California's natural landscape in the inexorable expansion of Los Angeles – a destruction that saw mountains and the Los Angeles river both built on and water drawn by aqueduct from the San Bernadino mountains to supply an ever growing population. At first glance, the parallels to be drawn with the expansion of Athens and the place of Kifissia within the greater Athens urban fabric and processes of development therein may appear to be few. Yet, the partial parallels, albeit that they exist at a micro-scale, are nevertheless there in more ways than simply the eclipse of the natural by the man-made environment but in the very agents involved in and processes by which the latter has been created.

Kifissia is located 15 km north of Athens city centre. It is the least populous, and has the smallest local economy of the post-suburban municipalities covered in this study but is one of the largest in terms of territory (Figure 4.1). From ancient times its natural environment has meant that it has always been considered a very attractive area. As one local history recounts,

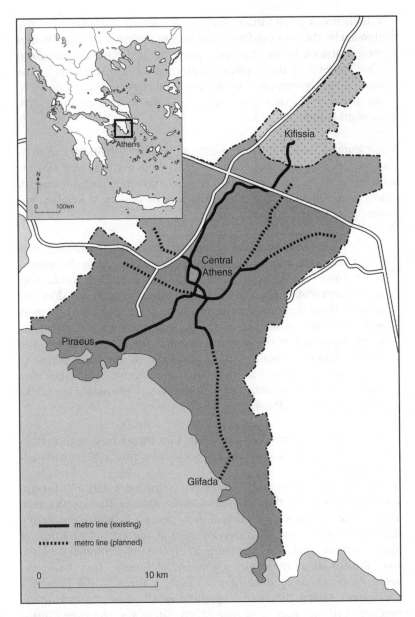

Figure 4.1 Map of Kifissia

Kifissia was a small village, built on a wooded and attractive area, flooded by the cool shadow of the platens and pine tree. The houses were scattered in the gardens, vineyards, and olive-groves, which were watered by the abundant waters of the Kifissos river.... The natural beauty, abundant waters and mild climate all contributed to the perception that Kifissia was one of the most beautiful resorts. (Simoni-Lioliou, 2002: 12)

These features were also the attraction to wealthy Athenians and over-seas tourists who had transformed Kifissia from a village to a sizeable summer resort by the early 1900s. A good example of rapid change of character of Kifissia at this time is the historic locality of Kefalari, just north of the Kifissia historical centre.

Kefalari did not have many permanent residents. The houses were few. There were a lot of harvest fields that belonged to the monas-teries. There were no neighbourhoods. The few residents would meet on Sundays and in festivals in church. Life in Kefalari was slow and simple. There was only one tavern named 'Kalabokas'. Then there were the hotels, 'Kourti's' and 'Apergi's' and there was some move-ment, especially in the summer months. Then the 'Secil' hotel of Costas Dimas was built. This was followed by the construction of 'Pentelikon' and later the other hotels in the centre of Kefalari were built. In this way, Kifissia started to sprawl slowly towards the north. (Simoni-Lioliou, 2002: 112)

However, it is from the 1960s onwards that increased urban develop-ment has produced the most marked socio-economic transformation of the municipality.

The 'spacious groves and melodious... plashing waters and tuneful birds' of Kifissia described by Gellius in the time of Hadrian (Karavia, 1988: 7) are a garden of Eden if not destroyed then considerably diminished by recent urban development at the expanding fringe of greater Athens area. The sheltering cool of Mount Pendeli and the river Kifissos made the resort settlement of Kifissia. They have also saved Athenians from drought from ancient times to the present – the waters slaking the near unquenchable thirst of the expanding, parched concrete mass of nearby Athens. However, the river Kifissos now lies buried beneath the concrete and tarmac of a major arterial route way out of Athens, the base of Mount Pendeli clawed at by sporadic housing development.

In studying urban development in Athens, Leontidou has highlighted some of the peculiarities of Mediterranean cities. For her 'urban phenomena which recur in Greece and all over the semi-peripheral world are usually considered as residual, culturally specific, even traditional or "precapitalist", destined to converge... with western patterns', however, 'spontaneous urban expansion is not a precapitalist remnant, nor a manifestation of residual peasant modes of land allocation. It has emerged with capitalist development and has been "functional" to it...' (Leontidou, 1990: 5). Perhaps the most peculiar feature of cities such as Athens then is the 'coexistence of "modernity" and informality' (Leontidou, 1990: 3). And it is this which presents the local variation on the post-suburban theme that is Kifissia, where older informal patterns and processes of development have shaped modern urban development at the fringe of Athens.

4.2 'Thinking Greek': A philosophy of urban development in Greece from the Republic to Metapolis

In the time of Plato, the city represented the territorial scale at which social cohesion manifested itself producing an archipelago of competing city-states. Particular city-states provided elements of Plato's ideal society and it was this territorial scale at which this utopia – *The Republic* – could be mapped out. Plato's *Republic* was both of and yet not of this world. From his home in ancient Athens, Plato argued in his 'theory of forms' that the search for knowledge was a search for super-sensible or eternal 'forms' and that the world of appearances could be the subject of opinion only (Russell, 1988). The logical property of Plato's forms, and in this we might include the universal form of 'city', stands in marked contrast to, for example, the world of appearances of the modern Athens. Modern Athens – 'metapolis' (Aesopos and Simeoforidis, 2001a and b) – is a settlement upon which we can have an opinion on the myriad pragmatic actions of individuals who, in 'thinking Greek', produced a particular rendering of the city as an ideal form.

Plato's theory of forms evinces something of the aims of this book. Implicit in this book is a desire to understand the city (and, by the same token, a nascent category 'post-suburbia') logically both as an idealised universal form *and*, at the same time, as so many imperfect representations of, or approximations to, that form. In this sense, post-suburbia is a form that currently is still being invested with meaning – 'an airy nothing' that in different contexts is attributed a 'local habitation and a name' in the terms of the opening quotation: here an 'edge city', there the 'outer suburb', 'technoburb' or 'suburban growth pole'. Plato believed that universal forms were god given. For the purposes of this book, Plato's metaphysics might be replaced by one that sees the most powerful of socially constructed essences achieve universal form in the guise of what have been described as 'world models' (Meyer *et al.*, 1997). For 'territory is not; it becomes... it is human beliefs and actions that give territory meaning' (Knight, 1982, quoted in Paasi, 1996: 32). Territories in the form of cities or post-suburban settlements have at once both a particular appearance – a local habitation and a name – and an ideal form that itself is subject to redefinition over time.

In arguing that contemporary urbanisation in southern Europe has been produced from distinctly capitalist processes, Leontidou is also clear regarding the uniqueness of southern Europe, and, by derivation we can suggest, the uniqueness of post-suburbia in the Greek setting. There are indeed distinctive patterns and processes of urban development that we will identify in this section but within these we can also detect aspects that, with differing intensities and at different geographical scales, parallel both the form and the processes of post-suburban development in North America, northern Europe and East Asia.

Informality and the impotence of planning

The history of planning in Greece may be divided into four periods (Christophilopoulos, 1997). The common experience across these four periods, however, has been the 'unsuccessful, inactive, fragmented and unrealised urban policies' which have led to the rise of small-holder real estate interests which function as 'points of resistance to attempts at reform, since they are an expression of the priority of personal gain for privileged groups over and against the general public interest' (Getimis, 1992: 252).

The first period (1828–1923) was characterised by the design of street plans for the biggest Greek cities without an appropriate legislative framework for urban planning. A second period – by far the most formative in terms of the sorts of urbanisation processes we are concerned with here – lasted from 1923 to the 1970s. The year 1923 was a defining year in the history of town planning in Greece, as this was the year when the parliament Act on 'Town, Villages and Conurbations planning and development' was introduced. In this context, land was classified into three categories: (a) areas within the 'city plan', which had an approved urban street plan; (b) conurbations that were created and existed before 1923 and had their own special legislative regime; (c) areas which were outside the approved street plans and outside the built-up areas that pre-existed to 1923 – the so-called areas that were 'outside the city plan'. Building activities 'outside the city plan' were regulated by ad hoc presidential decrees where building was, in practice, uncontrollable. The Act of 1923 has been modified many times, and there have been many deviations and distortions at the enforcement stage, leading to the development of residential areas which, despite having approved plans, lack communal space and related infrastructures. In addition, areas that were 'outside the city plan' have gradually been built up, either illegally or legally, by exploiting loopholes in the legislative framework. For instance, development outside the 'city plan' has been permitted provided that it is on the edge of national, regional or municipal roads, or even railways. Developers used these provisions resulting in 'ribbon development' at the edge of roads across the Greek countryside (Christophilopoulos, 1997: 96).

The years from 1970 to 1982 might be considered a third period. This era is characterised by the fall of the military dictatorship (1967–1974), the restoration of presidential parliamentary democracy and the subsequent production and ratification of a new Greek constitution which designated clearly the responsibilities and competencies of the State regarding urban development, the protection of natural and cultural environment. The ratification of the constitution was followed by a series of laws and parliamentary acts on 'regional planning and conservation' and the creation of a government ministry in charge of Environment, Public Works and Physical Planning.

Finally, a fourth period from 1982 to the present can be distinguished. One of the key developments in this era was the parliament Act of 1983[1] which increased the State's involvement in urban and town planning and aimed at tackling urgent housing problems. Although initially proposed as temporary legislation, it comprises the base of current urban planning. The 1983 Act led to the creation of state-led urban plans for all local government authorities in Greece, with subsequent legislation[2] permitting a measure of private urban planning activities to co-exist. A new law in 1997 introduced the principle of sustainability for the first time into plan formulation and made provision for the remediation and consolidation of loosely urbanised areas (OECD, 2004: 134).

Today, there is a situation in which most major cities have had plans that were finalised in the 1950s and 1960s but many smaller towns still do not have plans [Interview K1]. Since the early 1900s continuing to the present day an 'inherent problem of the current system is that legislation is created, modified and superseded but not actually implemented' (OECD, 2004: 135). Planning legislation typically has been weak allowing informal but rapid processes of land development to occur largely unhindered despite the growing powers of the state. 'The whole history of modern planning in Athens... is a history... of a class of landlords, proprietors and developers sabotaging planners and imposing unrestricted land transactions and building for profit' (Leontidou, 1990: 54). In part this stems from the highly generalised planning powers enacted since the 1920s, so that 'urban planning... remains general and undifferentiated, without adapting itself to local needs... this general and non-specific character of planning policy allows for the consistent exercise of individual pressure by interest groups' (Getimis, 1992: 244). Unlike North American edge city (Garreau, 1991) developments which embody a corporate-led expansion of the 'new frontier', urban development, as Leontidou suggests, amounted to a 'victory of popular control over the frontier of urban expansion after 1922' (Leontidou, 1990: 88). Despite significant attempts to strengthen the planning system in the most recent phase of legislative developments since the 1980s, this pattern continues to the present day. As one interviewee noted, 'everywhere in all respects you have piecemeal regulations penetrating what you think is a comprehensive process' [Interview K2]. These factors 'define a "non-planning situation" ' in which 'no consensus has ever been structured around planning... as has been the case in most countries in Europe' (Delladetsima and Leontidou, 1995: 284).

The State, Athens and the metropolitan region

The Greek state is one of the most, if not the most, centralised and interventionist within the EU (OECD, 2004). In particular, the competencies of local government authorities in Greece are determined to a great extent by legislation introduced by central government. 'The competencies of local government are designated on the basis of parliament legislation and not according to the National Constitution. This means that the central government has the ability, by introducing the appropriate legislation, to intervene in the activities of local government' (Patsouratis, 1994: 391). As a consequence, spatial planning has been a centralised function that only recently has become devolved to lower tiers of government (OECD, 2004: 132). 'The current situation... is characterised by a highly centralized administrative planning system; a weak local government structure with disjointed and loosely defined responsibilities... [and] an immense number of public and other institutions which directly or indirectly influence through their actions spatial development' (Delladetsima and Leontidou, 1995: 284). Moreover, within this complicated hierarchical institutional setting, party-political channels and clientelistic relations dominate as means of exerting influence. As a result, 'local government continues to develop vertical party-dominated relations of hierarchy and dependence with the state pursuing resources and clearly defined competences' while more generally the 'gap between the weak civil society and the strong state is filled by the emergence of clientelistic relations which function as the main informal channels of political integration and participation of society in the public administration and political system' (Getimis and Grigoriadou, 2004: 12).

The competencies of various institutions pertaining to urban and regional planning are specified by a wide range of laws and presidential decrees, but the general principles underpinning all legislation have been laid out in the Greek Constitution of 1975. According to Article 24 of the Constitution, the Greek State has the obligation to take all the necessary measures for the protection of the natural and cultural environment. The Greek State has the obligation and the exclusive authority to control, regulate and direct urban and regional planning and restructuring.

Since the ratification of the Greek Constitution of 1975 there have been a number of more specific laws pertaining to the competencies of various organisations and institutions. The government Ministry of Environment, Public Works and Physical Planning has the exclusive authority to design and implement urban and regional policies and to design and approve town and city master plans. The ministry has representatives at the regional level: there are 13 regional directorates based in the administrative centres and they supervise activities that are directed by the ministry. It can therefore be argued that the activities are much centralised. At the local level, municipalities are the first point of contact for issues pertaining to planning applications. Municipalities would then pursue the approval of the ministry and if this is granted they would be in charge of implementing any plans or carrying out the appropriate actions (Christophilopoulos, 1997).

Growth in the Greek urban system is highly focused on just a few major urban centres, most notably Athens but also Thessaloniki. Under laws enacted in 1985 these two cities made provision for semi-autonomous, non-elected bodies heavily supervised by central government ministry to create guidelines for metropolitan development [Interview K2] (Gerardi, 1997). In particular, the 'Organisation of Regulating Plan and Environment Protection of Greater Athens' was founded[3] as a 'body of zoning and urban planning, protection of the environment and programming' (Gerardi, 1997: 243). Despite the primacy of a city like Athens and its condensing of urban-industrial and population growth within Greece, there is an absence of strategic planning at the metropolitan scale.

> The concept of metropolitan municipality is very important in other European cities, but not here... You could say that the idea of the Metropolitan Municipality has not yet matured politically. Perhaps the ministry [Environment, Public Works and Physical Planning] in charge did not deal with this issue as it should have. We do not know anything about possible objections of municipalities, because the project was never put on the table. It has never been formally discussed. There were some expressions of objection in private conversations that we had with mayors of regional municipalities, but there was no formal, serious discussion. [Interview K12]

Indeed it was suggested by another interviewee that those attempts to institute metropolitan scale planning arrangements have effectively been eroded due to pressure from local authorities who in turn convey broader local social pressures [Interview K2].

After 1994, municipalities have managed to obtain certain competencies from prefectural level. Some municipalities have pushed for this in order to gain greater control on developments in their areas but many are not bothered and they can only assume competences by permission of central state [Interview K1]. So, for example,

> Some big municipalities have the ability and technical support and accessibility to centres of political power to a greater extent than others and they have the ability to negotiate with political actors.... Most of these municipalities are in Athens... This depends to a certain degree on the size of the municipality but also on the personality of the mayor. [Interview K2]

Whilst local governments have in some instances been able to gain powers in relation to controlling urban development more effectively, the combination of a centralised state and the primacy of Athens within the national economy have gradually exerted increased development pressure in peripheral metropolitan municipalities such as Kifissia.

> The municipality of Athens sees the growth of the peripheral municipalities as a good thing. We feel that Athens is

overcrowded... Generally we want Athens to be decongested and one way of doing this is to further develop the suburbs. For instance, Kifissia is a good case, it has many advantages... [Interview K12]

Whilst anti-growth sentiments have already emerged in Kifissia and have prompted the assertion of powers to prevent further urban development, these bear the weight of pressures for decongestion in the Athens metropolitan region.

Thinking Greek: Polykatoikia, densification and sprawl

As we have seen above, the main structural elements of the rapid and generally unplanned urbanisation of the post-war era have been the 'non-planning' situation in which there has been a fragmented and informally organised pattern of private landownership and development on the one hand, and the condensation of development pressure upon the capital city and through its central state institutions, on the other. It is these dynamics that have, in the first instance, produced the dense fabric of modern central Athens and latterly a similar densification of the outer suburbs such as Kifissia.

Within this structural context, agency in the form of 'thinking Greek' has ensured that 'the contemporary Greek city presents characteristics different from Western-European or North American cities, which are designed and planned to a smaller or larger extent; it's formless, borderless and placeless urban landscape subjugates any aesthetic' (Aesopos and Simeoforidis, 2001b: 32). Instead, 'public space is strongly related to the resolution of one's personal problems by one's own initiatives' (Getimis and Grigoriadou, 2004: 13). In the most practical terms, this reduces itself to the possibility of building something and seeking solutions to subsequent problems (such as securing the appropriate regulatory approval or licenses).

We will return to the essentially unplanned nature of urban growth below but before we do so the origins of the 'formless, borderless and placeless urban landscape' of Athens and other Greek cities must be traced to the 'antiparochi' system of dwelling construction.

The boom of the Greek economy from the mid-1950s onwards led large numbers of the rural population to migrate to the large urban centres, primarily Athens. The need to house these people was met by the numerous 'poly-katoikias' (apartment buildings) constructed by small or medium sized contractors operating on the 'antiparochi' ('quid-pro-quo') system of exchanging land for built space. The 'polykatoikia' in its endless proliferation, gave shape to entire Greek cities. (Aesopos and Simeoforidis, 2001a: 21)

In the words of one interviewee the antiparochi system ensures that 'nobody loses' in that one or more apartments will be handed over to land owners in return for the right to build on their land [Interview K1]. Despite the informal nature of this development process and despite the dominance of small- and medium-sized construction companies in numerous small-scale developments, the development process is never-theless regarded as every bit as capitalist (Leontidou, 1990; Papamichos, 2001) as the larger scale corporate developments of post-suburbia found elsewhere in Europe and in North America. The antiparochi relation 'in spite of the fact that it makes its appearance in the archaic form of barter . . . is to be understood . . . as a relation of purely capitalist content' (Papamichos, 2001: 82).

In central Athens, development was typically accomplished by building vertically and adding floors to existing building blocks, a process which in itself appears to have driven the perfection of a building form – the 'Polykatoikia' – that, in its simplicity and robust adaptability, Le Corbusia would surely have approved of.

> The Greek polykatoikia is . . . a building type that offers construc-tion simplicity, economy and durability, a prototype to be repeated to infinity. . . . This is the pragmatic response to a real issue the upgrading of the quality of life through a mass-yet private-housing system. (Aesopos and Simeoforidis, 2001b: 33)

Elsewhere, in the available land between buildings in less densely urbanised parcels of land towards the edge of major cities, entire new apartment blocks would be erected, though here again the polykatoikia would be the preferred form. Through this piecemeal process,

> The Greek city constitutes the over ambitious transformation of a relatively small whole into an urban hyper concentration through the continuous repetition of a unit, a process with minimum organ-ization or programming, based on the microscale: the polykatoikia and the small- to medium-size contractor. This private urbanisation process is implemented through the 'antiparochi' system (exchange of land for apartment surface) . . . The polykatoikia is at the same time the infrastructure . . . and the superstructure. (Aesopos and Simeoforidis, 2001b: 37)

What we see here is the proliferation of an urban form, every bit as potentially alienating as archetypal Los Angeles – albeit at a different geographical scale. If Los Angeles stretches horizontally with repeated differentiation of land parcels in a confusing landscape (Dear and Flusty, 1998), the density of central Athens reveals repeated differentiation in use vertically and in the small spaces between blocks. As such, 'the Greek city is continuous. The immense layer of equal depth of the built indifferently covers the natural topography' and 'with few exceptions . . . public spaces are the residue of the built' (Aesopos and Simeoforidis, 2001b: 48 and 45 respectively).

What might at first glance appear to be unique to Athens is in fact, when we recall observations on the growth of Los Angeles, perhaps a feature more common than hitherto admitted of cities east, west, north and south. In arguing that the seemingly free-market sprawl of Los Angeles was planned, Hise (1997: 52) draws attention to the fact that such planning existed at the micro- or neighbourhood and the macro- or regional scale but not the crucial integrative meso-scale of the city. In this particular respect – the absence of an integrating meso-scale of planning – there are some similarities with Athens. So, for example, 'there is no mediation between the polykatoikia and the city. Beyond the polykatoikia, there is Athens, no formal hierarchy in between' (Sarkis, 2001: 155). In this, and in the descriptions of Athens above, there is something familiarly 'confusing' about the visual appearance of the built form, albeit represented on a micro-scale when compared to Los Angeles' macro-scale.

As a result of this pattern of development, 'the urban environment in Greece today reflects the distance that exists between the perceptions of planners and decision makers and the public' (Amourgis, 2001: 77). This 'self-financed private property pattern and the expansionist logic deriving from it' has resulted in the legalisation and incorporation of more and more peripheral areas within city or town plan boundaries (Delladetsima and Leontidou, 1995: 283–284). This remains the situation and provides an accurate caricature of the development process driving urban expansion in Greece and especially Athens. It also sets the context within which developments in Kifissia at the edge of Athens can be understood. Essentially, as one interviewee noted, 'the planning follows the existing situation' [Interview K1]. With the densification of central Athens reaching its limits this situation applies most clearly at the edge of major cities such as Athens. Here, then, in 'the face of anarchic urban sprawl, planning policies were then developed in a fragmented manner: measures were introduced ex post to legitimise existing squatter settlements and to adapt to the interests of landed property, and especially to pressures from the owners of small land parcels' (Getimis, 1992: 243). This is something that we go on to explore in the case of Kifissia in the following sections of this chapter.

4.3   Kifissia: Playground of the Athenians

The origins of Kifissia's initial major expansion into a suburb of sorts followed the pattern of the numerous streetcar and railway suburbs associated with major North American and European cities (see Figure 4.2). The migration to this particular suburb was a select one indeed as one interviewee intimated.

> Since 1850, as in all of Europe . . . Kifissia developed because we had the train. So when the train from Athens reached Kifissia . . . the rich people came and built the big houses in a way copying the English, French or German type of life. And that's why . . . very near the Railway station and near to Kifissia Avenue you will find all types of buildings. Neo-gothic, neo-classical, French-style of castles . . . all sorts of architectural styles. [Interview K3]

However, from the outset, Kifissia was never as mono-functional as many of the North American and European suburbs. As the quotation above implies, this was an extremely wealthy suburb in which many of the villas were essentially summer residences. Moreover, it became home to a number of major hotels and hence might be more accurately regarded as a resort town for wealthy Athenians and international tourists. Kifissia remains a very high income suburb not only within the Athens metropolitan area but also within the national context [Interview K13]. However, further changes in its physical appearance and function have seen it evolve into a more fully post-suburban municipality.

Until the last two decades of the twentieth century, Kifissia remained a relatively exclusive suburb. Relaxation of building regulations under the dictatorship (1967–1974) had the effect of intensifying the urban fabric both in promoting the growth of second homes and building on very small subdivisions and the addition of storeys to older buildings [Interview K1]. As one interviewee described,

> Kifissia is a very nice suburb. It used to be a summer resort. And the very high income classes have traditionally resided in Kifissia. In the last few years though Kifissia tends to lose its beauty and its identity. It is a very nice city, an aristocratic lady in decline, because of the cement. [Interview K4]

Figure 4.2 'Old' Kifissia

This view of 'Kifissia in decline' seems to be shared by a large number of local people and it is indicative that one of the few books on Kifissia is written in a style that expresses a sense of permanent and irrecoverable decline (Simoni-Lioliou, 2002).

On the one hand, and at a distance from the centre of Kifissia, the general outward expansion of Athens and the growth in manufacturing and other industry in Greece saw the extension of larger industrial, commercial and retail enterprises along arterial roads. One of these – the national highway – runs through the west of the municipality of Kifissia. As a result, Kifissia is also now home to a number of major multinational companies including Alstom, Hoya Lens, Metaxa, Tria Epsilon (licensed bottling for Coca-Cola in the Balkans). Under Greek law, land owners give a portion of their land for public works such as roads but fragmentation of landownership means that this is often insufficient to generate coherent and consistent zones of land use [Interview K1]. The scene along the national highway in Kifissia's territory is thus rather chaotic. These major firms huddle together along either side of the highway to create a linear industrial zone rarely more than one factory deep (Figure 4.3). Behind, poorly surfaced roads weave unevenly among fields to houses and, in some cases, are the only major access to factories (Figure 4.4).

Figure 4.3   Companies along the national highway

Figure 4.4   Land-use mixing behind the national highway

On the other hand, and independent of the reach of the most obvious tentacles of expanding Athens, the central area of Kifissia has also been affected by an intensification of the urban fabric. 'The permissive building code introduced during the dictatorship was taken advantage of, and affected all urban areas except for a few bourgeois enclaves... which resisted it' (Leontidou, 1990: 220). Kifissia might be considered as one such bourgeois enclave but even here the effects of the changes in building codes were felt.

> After the 1970s... the building regulations changed and all the land owners could build higher buildings and they started mixing functions. And we had either houses or small shops. Steadily the character of Kifissia changed from a residential area to a rather mixed area. [Interview K3]

The first trend to note relates to the intensification of residential land uses within Kifissia. Starting in the 1960s some of the large villas built in the 1800s in the initial growth of Kifissia as a suburb began to be demolished or their land subdivided in order to build apartment complexes for those wanting permanent residences away from central Athens [Interviews K5, K6]. The owners of these villas appear to have been willing participants in the process at this time though opinion has subsequently changed [Interview K5]. In the most extreme cases, modern apartment blocks of five storeys sit adjacent to the two-storey villas of old Kifissia. Here the special character of Kifissia has to a degree lent itself to piecemeal intensification and mixing of uses. Kifissia has 163 listed buildings – a portion of them remaining in poor condition due to the cost of maintenance [Interview K3]. The better appointed villas have been converted into offices, the land of others taken for new developments and a fraction of the remainder steadily decaying.

A second aspect of Kifissia's changing character to be drawn from the quotation above is that of the mixing of land uses and functions. The fragmented pattern of urbanisation produced from the small-scale, privatised but nonetheless capitalist development process typical in Greece has produced 'a patchwork of economic activity and social classes throughout the urban fabric... [that] contrasts with zoning of economic activity and segregation of social classes in the North' (Leontidou 1990: 12). This process began somewhat later than the intensification of residential land use. Beginning in the early 1980s, the gradual but persistent piecemeal developments have also seen Kifissia expand into a significant retail concentration within the greater Athens area

(OECD, 2004: 100) and something of a retail and nightlife playground for Athenians. As one interviewee described of the changing land use of central Kifissia:

> I remember... 20 years ago, I was looking at Kifissias Avenue and there was nothing built between Psyhico and Kifissia... And then in the 1990s big buildings started being built and now it is full of buildings. The area between Psyhico and Kifissia, the land had no great value... Now, everything is like a big city. [Interview K7]

This building is composed of offices and retail uses along and off of Kifissias Avenue – the main road through the municipality – and can be traced to the early 1980s prior to a tightening of laws governing land uses in Kifissia.

> The big problem started in the 1980s when they made the first shopping centre of Kifissia... But at the time nobody could understand that they were opening the bag of Aeolus.... When this was built it was a big success and it whetted the appetite for all the developers to do the same thing. After that we had five or six shopping centres that were developed around it until we had the time to change the laws but they had their permissions and they were there. That is why if you go to the centre of Kifissia it is a terrible mess. [Interview K5][4]

Furthermore, some establishments have been built and operate without the appropriate permits but have not been shut down [Interviews K3, K8]. Indeed one interviewee suggested that complaints by the public to the municipality against businesses operating without appropriate licences had been met by threats [Interview K3]. The enforcement that does take place is often ineffectual since fines are small in comparison to the revenues of the businesses concerned [Interview K8]. In sum, as one interviewee confirmed,

> Kifissia has always been a very favourite suburb of Athens because of the climatic conditions... After the 1980s all the wealthy Athenians... used to... invest here, to live or to work or to establish an enterprise. So gradually from an absolutely residential area... it has become a centre for economic activity. [Interview K8]

The growth of shops, bars and restaurants in the central area of Kefalari ensures that the population of Kifissia expands and its character changes markedly at the weekends. From the slightly exaggerated perception of a representative of the Association for the Protection of Kifissia, 'during Friday and Saturday we have at least 5–10,000 people rushing to go to the bars or to the restaurants. Never visit Kifissia on Friday or Saturday near Kefalari, you will think you are in Las Vegas!' [Interview K3]. Undoubtedly, Kifissia is not Las Vegas. However, as Figure 4.5 illustrates,

Figure 4.5 Retail development in Kefalari

the retail-therapeutic environments of Kifissia do nevertheless transport one to some exclusive suburban California address.

The further intensification of the economic function of Kifissia during the last two decades of the twentieth century has also been driven by the prestige that a Kifissia address can confer upon businesses. Several of the old landmark hotels and larger villas of Kifissia have been renovated to form prestigious offices for banks and, paradoxically, shipping companies operating from the port of Piraeas. Perhaps the most notable of these is the headquarters of the Greek multinational bank, Eurobank. In the early 1980s, the Latsis Group that owns Eurobank went to the central government Ministry of Environment, Public Works and Physical Planning to obtain a special law enabling a change of use of a major hotel into office space – by-passing the municipality. The refurbishment of an old hotel into office space by the Latsis group was not especially controversial given the sympathetic refurbishment and the provision made for parking space [Interview K5]. However, the case of Eurobank also highlights the gradual shift in opinion and policy towards conservation after two decades or so of urban development in Kifissia. Having refurbished an old hotel in central Kifissia, a subsequent application to build a modern extension to these premises was recently refused after a 3-year court battle fought by the Association for the Protection of Kifissia [Interview K3]. Other such refurbishments have been more controversial. Eurobank's offices are on one side of Kefalari square. On another side is the Semiramis Hotel. No change of use has been made to the building but the style of refurbishment was approved by an architectural committee not at the municipal level but at the higher Attica prefecture administration. Semiramis' luminous pink, and green modern trim, has been derided as out of keeping with the traditional architecture of Kifissia and the rest of the square (Vardas, 2004).

Further and perhaps more striking examples of this mixing of land uses can be seen elsewhere in the municipality. Taken as a whole, Kifissia as a municipality clearly no longer has a single identity or function. But, more fundamentally, this fragmentation has resulted in a mixing of land uses and functions in a manner not dissimilar in type though most definitely dissimilar in geographic scale and economic intensity to that highlighted in south-east Asia. So, for example,

Near the national road near Coca Cola – it's still Kifissia – Adames. It's a place where it's really undeveloped. It was very much damaged by the earthquake of 1999. This area has been characterised as an industrial park... but if you visit there, there are streets that are narrow. . . There's a man who lives in a one storey house. Then there's a two storey small factory. And now they have decided all this area must function as an industrial park. I don't think they will ever be able to succeed in this function because the local people they always react. The politicians always retreat, they never insist on their decisions. [Interview K3]

4.4 Post-suburban growing pains: Conservation versus growth

As we will see in the case of the urban entrepreneurialism apparent in Getafe, mayors can play a pivotal role in shaping the character and pace of urban development. Despite the special character of Kifissia and subsequent pressures for conservation, successive mayors in Kifissia have been generally permissive of a pattern of intensification of the urban fabric typical in Greece as a whole. Moreover, strategic leadership of the sort found in Getafe has been conspicuous by its absence in Kifissia. So, for example, Mayoral leadership in Kifissia pales in comparison to that found elsewhere and notably in the neighbouring municipality of Maroussi. Moreover, as one interviewee commented,

I think some municipalities are much more dynamic than Kifissia. For example, our neighbouring municipality, Maroussi . . . the Mayor . . . we always come back to the Mayor. . . . He's very well connected. He's also the same political party as the ruling party now in Greece. In a way his municipality has been chosen as a pilot for all Greece. But he operates with very private standards. He is totally different from our Mayor. He runs a multinational company. [Interview K8]

Thus, although the character of Kifissia has changed quite markedly from the 1980s it does not resemble the wholesale development that has occurred in neighbouring Maroussi – commonly referred to as 'Vovopolis' after a property developer responsible for several large-scale routine office and retail developments in the municipality. One interviewee described how these developments were of a different scale to those commonly found in the US.

This kind of development in Greece does not necessarily occur in the edge of the city. The name par excellence in the Greek case is Vovos. . . . He builds inside, beside main roads like Kifissias Avenue.

However, with the scale that you compare it, it is still small scale. [Interview K2]

No single property developer has been able to exert such influence in Kifissia. Rather its contemporary appearance has been the product of many small-scale property developments and conversions and lack of enforcement. The spectre of 'Vovopolis' of nearby Maroussi is something that has been keenly resisted in Kifissia. For fear of the negative externalities, the municipality and business and conservation interests alike have been content to remain at some remove from the construction and infrastructure improvements for the recent Athens Olympics – much of which have been carried out in or will benefit neighbouring Maroussi. Further large-scale developments promised by the Mayor of Maroussi, such as a proposed theme park in the forest area of Vrilissia at the border of the two municipalities, have been opposed by conservation interests.

The lack of major shifts in strategic direction associated with mayoral leadership also affect the overall complexion of political representation on the council in that perhaps only a fraction of the council members have changed with a change in mayor. Issues of corruption and patronage in the local development and regulatory process might be seen as both cause and effect of such stagnation in political representation.

Present day Kifissia presents a familiar post-suburban tension between pressures for conservation on the one hand and further economic growth on the other. Further degeneration of the municipality's abilities to control development has been halted. After several relaxations of building regulations during the past three decades or so, the latest law requires that no new building regulations can be applied that are less stringent than those that currently exist [Interview K3]. Only since 1991 has Kifissia had a town planning department with the capabilities to regulate development. Since this time they have issued several regulations relating to the prevention of further mixing of land uses. As one interviewee noted,

The trend now is to control development so as not to change the character of the town because Kifissia it offers a very nice environment of all the suburbs of Athens. So they don't want to lose that. But at the same time, for example, Maroussi, which is very close it's also close to the Olympic stadium and it will benefit a lot from the Olympic games and it's the neighbouring municipality and they see that they don't want to lose the train of the Olympic games. [Interview K8]

However, the problem of combating 'Greek thinking' – a problem of implementation and enforcement – remains [Interview K3].

Opinions vary over the precise balance of political and popular forces in favour of development and conservation. As a result of the intensified development of Kifissia that began in the early 1980s, the political complexion of the municipality has changed over the last decade with the election of conservative mayors from the Nea Dimokratia party. As one interviewee noted, 'Kifissia used to have strong Communist Party presence... But with the development of Kifissia, gradually the socio-economic structure of the population and the voters has changed' [Interview K13]. With this change in socio-economic and political complexion, conservation interest groups have begun to have greater influence in municipal politics.

> Since then there have been many variations because these laws have been changed quite a lot but always in the direction of preservation. This has happened because we have here not only the municipality of Kifissia which was in favour of these measures but also local associations which were pressing for this matter. On the other hand are the developers and finance companies that are pushing from the other side. And there was a strong battle at the time in the city. Sometimes they get to get their way.... And of course, if you lose a battle it doesn't mean you lose the war. But then after you lose another battle and another battle you cannot regain what you have lost. [Interview K5]

The same interviewee estimated that as much as 85 per cent of the voting population of Kifissia were now in favour of conservation. The picture is more divided according to another interviewee,

> There is this problem with half the people living here they want Kifissia to a have a residential character and half the people they want development, they want companies to come here. So there is a conflict on this subject. Because they still want to keep the residential character because the value of the land is extremely high here. [Interview K8]

What is not in doubt is that local politicians themselves appear open to persuasion on the issue in the sense that local pressures from constituents are responded to on the basis of garnering votes for re-election [Interviews K4, K5].

In part, this tension is also structured by possibly unusual labour market and residential dynamics throughout the greater Athens region. For example, few of the major employers in Kifissia draw their labour from the local area. The local dependence of a company like Tria Epsilon located on the National Highway is minimal as an interviewee described 'when this factory was founded 30 years ago there was only farmland around here, so historically the labour force was pooled from all over Attica. I wouldn't say that we have any particular relationship with the Kifissia labour market' [Interview K9]. Moreover, problems of unemployment and social exclusion in Kifissia have been defined in quite unusual terms. With the growth of its retail and entertainment sectors, Kifissia is an importer of less-skilled labour. However, it has been argued that there is a lack of local employment opportunity for Kifissia's generally highly educated residential workers who are reluctant to travel longer distances to work elsewhere in Athens [Interview K8].

Moreover, a degree of disconnection between residence and local politics uncommon in many other nations is apparent in Greece and the Athens area. In particular, Greeks typically register and enrol in the electoral catalogues of the municipality where they were born and tend to stay registered in the same area, even if they move out throughout their lives. There is even a government term for this type of voter: 'heterodemotes' – those who come from and still are citizens (demotes) of another (hetero) municipality and still vote for members of parliament, mayors and councillors in that area. Voting is a right as well as an obligation according to Greek law – it being illegal not to exercise that right. Up to 1998 the estimated 1,500,000 'heterodemotes' (over 10 per cent of the Greek population) would travel *en masse* on election day, using specially chartered flights or coaches, or by boat or train and they would purchase a special election-discounted ticket (Sokos, 1998).[5] The implication of this arrangement is that voters who do not live in the area anymore may have considerable influence upon the outcome of local elections.

4.5   The subterranean politics of growth in Kifissia

In Greece, in general, the extent of public–private partnerships and the role of the private sector within them has been markedly curtailed by the sorts of central state dominance and party-political mechanisms of interest mediation described earlier (Getimis and Grigoriadou, 2004: 17). In particular, the pattern of business interest representation reflects this general pattern centred on party politics and mayoral patronage.

As one interviewee described, 'unfortunately we have not reached the level at which these actors would try to affect decisions in a structured way... we end up doing personal lobbying – even if there is no need they are trying to influence in a favouristic manner' [Interview K2].

Following the national pattern, collectively organised business interests are virtually non-existent at the local municipal level. There is also a parallel absence of formal channels through which business interests influence decision-making at the municipal scale [Interview K13]. The larger firms along the National Highway in the western reaches of Kifissia have been lobbying the municipality. However, as one interviewee observed rather wryly, 'they are organized by themselves, they are organized individually so the pressure can be handled!' [Interview K5]. In the case of the industries by the main highway, their sheer diversity has until recently tended to militate against collective interest representation [Interview K10]. They have also recently lobbied central government through an informal and rather ad hoc collective grouping. Nevertheless, one interviewee suggested that these companies had begun to organise themselves on a more formal collective basis [Interview K11]. They have also been lobbying the municipality. This appears to be a response primarily to concerns over the condition of local roads providing access to and from the highway as detailed earlier. The many shops, restaurants and bars in central Kifissia are not organised into any collective business body. Indeed, as one interviewee commented, 'the big commercial chain shops do not want collective bodies... they prefer to act individually because they have the financial standing and power to do so' [Interview K4]. The only formally constituted collective business body in existence in Kifissia consists of a grouping of businesses in a disadvantaged retail sub-centre to the east of central Kifissia – Alonia. The Commercial Association of Alonia was first organised in 2003 in an attempt to get the municipality to provide environmental improvements in an area generally neglected in favour of the larger and more vibrant central area [Interview K4]. Elsewhere, business people were aware neither of any significant local collective business interest representative bodies [Interviews K1, K7, K8, 10] nor of any great interest on the part of the mayor or the municipality in the health or problems of the local business community [Interviews K4, K7, K10].

This description highlights something quite fundamental about the pattern of business representation in Greece – namely the large-scale absence of local business organisation and representation and the channelling of business interests either indirectly via the local mayor or

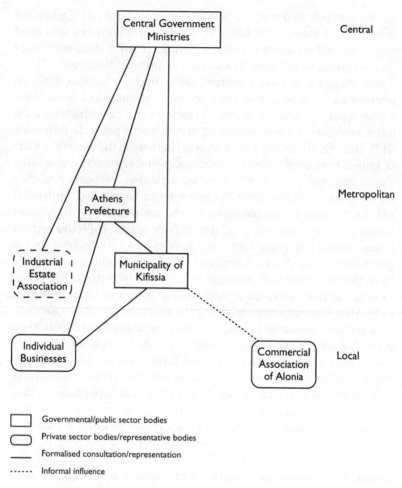

Figure 4.6 Map of business interest representation in Kifissia

politicians or directly to the relevant central government politicians and
ministries as depicted in Figure 4.6. As one interviewee highlighted,

> if you are affiliated to the political majority in the council they
> will help you – if you are affiliated to the opposition they will
> not.... Local politicians are voted for on partisan basis and they are
> not judged from their actions and results. Everything is partisan –
> there are no autonomous movements that would promote the

solution of the problems that this place needs to solve. The local government in Kifissia is limping. Bad management, bankrupt firms, over-recruitment of redundant personnel. [Interview K4]

There is a significant collective organisation representing environmental and urban conservation interests – the Association for the Protection of Kifissia – which appears to have carved out some system of formal input into municipal policy-making albeit through a planning advisory council [Interview K3]. However, in general, it would be fair to say that interest representation is not systematised and hence this feeds through into the sort of laissez-faire approach that has characterised Kifissia's development since its heyday as a resort town and exclusive suburb.

In sum, personal connections, often channelled along party-political lines, are the major form of getting things done. This process works at the local level and upwards and both for and against economic interests. It has been a major mechanism allowing urban development in that favoured developers are invited to present their proposals to planning committees [Interview K3]; bar or restaurant owners are allowed to continue operating without licenses or repeatedly infringe them [Interviews K3, K4, K8]; but also some of the recent local building regulations issued since the 1990s have been forged through personal connections at central ministries [Interview K3].

4.6 Conclusion

Whilst relatively small in scale in comparison to the other post-suburban municipalities considered here, Kifissia has nevertheless undergone a distinct transformation from a resort town and residential suburb to an employment and leisure centre of some significance within the greater Athens metropolitan region.

Kifissia's pattern of post-suburban development reveals a degree of informality not found in our other southern European case-study municipality, let alone those drawn from northern Europe. This informality is evident in the very process of development itself which has been highly fragmented and individualised to the extent that no major developers have dominated in Kifissia's urban expansion and socio-economic transformation. On the one hand, while the actors involved – residents themselves alongside small-scale building and development companies – differ from those commonly dominating in northern Europe and North America the 'intentionality' driving post-suburban development has been capitalist in nature. Then again, on the other hand, in the aggregation of so many micro-planned, discrete developments, the apparent non-planned gradual 'post-suburbanisation' of Kifissia again evinces some similarities with post-suburban experience in North America – albeit at a different geographical scale.

Moreover, Kifissia provides a European illustration of one of the post-suburban tensions raised in Chapter 2. In contrast to the case of Getafe where there was a balancing act played out quite explicitly in urban politics over the attraction of employment and the development of collective consumption expenditures, informally organised urban politics has instead focused on the tension between promoting further growth and urban development versus conservation of the urban environment.

Unlike institutions in its southern European neighbour, Getafe, which we consider next, local government in Kifissia has not been effective in enlarging its spaces of engagement to capture important capacities and expenditures from higher tiers of government. Instead it is private-sector institutions – property developers and builders and even individual residents, through political patronage – that have been able to mobilise support for their particular agendas external to the municipality.

The distinctive process of development in Kifissia and its fragmented nature appear also to have produced a quite distinctive post-suburban form featuring, generally, a low-density, low-rise and porous-mixed land-use pattern of development quite unique among the cases considered in this book.

5
Getafe: Capital of the Gran Sur

There are men and women
who know what to cling to
Taking full advantage of the sun
and the lunar eclipses too
casting aside the useless
but making full use of what is valuable
Thanks to their ancient faith
the South has its existence.

Translated from Mario Benedetti,
'El sur también existe',
Preguntas al Azar

5.1 Introduction

In contrast to Noisy-le-Grand, although a post-suburban *space* imposed
by the central state, from an early point in this process, local institutions
and more importantly significant individuals such as the local mayor
have been able to construct Getafe as a distinct post-suburban *place* –
a place with distinctive social and political concerns from which to
enlarge their spaces of engagement within the wider metropolitan area.
The stanza from Benedetti's poem above is one that the mayor of Getafe
himself has been fond of quoting and indicates something of the agency
that has seen Getafe emerge as an invented space from the disadvantaged
south to challenge Madrid.

While the postmodernity of Los Angeles (Soja, 2000) is rarely to be found
to the same degree among European cities, the Madrid of Almodóvar's
films is considered to have been portrayed as a city of fragments.

Madrid's cityscape is a mix of diverse spectacles such as a city centre characterised by a net of avenues... suburbs marked by a disorderly development; ultra-modern glass and steel skyscrapers along the Paseo de la castellana; and working class houses in pink or red in the southern, industrial areas. Different urban landscapes and opposing lifestyles are sometimes post-modernly contiguous. (Mazierska and Rascaroli, 2003: 33)

There are other deep-rooted respects in which Madrid might be considered a postmodern city as the same authors note. Madrid was itself imposed as the national capital by Philip II in 1561 without being a natural choice based upon its population or economic role when compared to cities such Salamanca or Toledo. With a population of just 16,000, it was a fraction of the size of other European capital cities, such as Paris or London, at the time, and was for several centuries afterwards referred to as 'solo Corte' – just the (Royal) court (Santos *et al.*, 2000: 159). Madrid has only gradually assumed some social and economic significance as a European capital ostensibly from the migration of population from the rest of Spain.

It was only in the first decades of the twentieth century that Madrid began to assume importance as a financial, commercial and communications centre to emerge as a modern capital city within Spain. It was also at this time that a distinction between the city's core and its peripheries first began to emerge with the novelist Pío Baroja describing the city as having 'a refined almost European life in the centre... an African one in the suburbs' (quoted in Santos *et al.*, 2000: 428).

Moreover, historically Madrid's imposition as the capital city 'has required reinforcement with regard to communications with the rest of the country.... This demand for radial communication has had a major influence on the structuring of the city growth' (Maldonado, 2002: 361). And indeed, Getafe appears to owe some measure of its growth to its accessibility and as a point of entry to the capital city-region [Interview G10]. As one interviewee described,

Getafe has always been a point that has been travelled through, going back to the late middle ages. It has never been an isolated place, and in the modern era it has had the aero-nautical industry which is also about communication, and in a way this spirit of good communication explains certain things today. [Interview G1]

Notwithstanding such a heroic appeal to history, Getafe was, rather like its capital city neighbour, ostensibly an invention of the state, albeit a much more recent invention. In the immediate post-war years, there was a concerted effort made to further industrialise Spain through the development of a greater Madrid. Getafe, as one of the most accessible existing settlements within the wider metropolitan region, was chosen as a major site to house new industry and workers. Getafe therefore grew from a small satellite town of Madrid with a population of 12,500 in 1950 to 150,432 by 2001. Its population doubled in the 1950s and tripled in the 1960s, after which time growth has slowed somewhat (Santos *et al.*, 2000: 547).

Getafe is therefore an important southern industrial fragment of the postmodern mosaic of Madrid. Indeed there is a case for suggesting that locally based institutions and actors have created Getafe as a place – a place with some assertiveness within the metropolitan space of Spain's invented capital city.

5.2 The cities of the plains: El reino de taifas

During the boom in the Spanish economy of the 1960s, Madrid benefited the most in terms of employment. Almost 700,000 migrants came to the city between 1960 and 1970. Its population rose from 2.4 million in 1960 to 3.6 million in 1970, and around 40 per cent of all housing units in the Madrid metropolitan area in 1975 had been built after 1960 (Castells, 1983: 220). And so, the small towns and villages which surrounded the capital – particularly in the south – were over-whelmed by waves of rural migrants. Getafe was no exception.

Indeed in some ways Getafe epitomised this phase of expansion. In the early post-war years it was selected as a key site for Spain's late drive to Fordist industrialisation (Holman, 1996). Central government decided that the state aviation engineering company La Factoria Construcciones Aeronáuticas SA (CASA) should locate in Getafe where there was also a major military air base. Shortly after, in the 1950s, three major multina-tional companies – Kelvinator, John Deere and Siemens – arrived. These four companies formed the industrial heart of early expanding Getafe. 'These four have generated a little micro industry if you like of metal and machinery' [Interview G2]. CASA especially has conferred a particular industrial complexion on Getafe and continues to do so today. CASA, for instance, dominates as a client of public-sector training and techno-logy institutions such as the Fundación de Innovación. As the director of the latter described, 'here in Getafe what is different from the rest of the municipalities in the South is CASA which needs a lot of providers – a lot of small enterprises – working for them and they are especially localised here in Getafe' [Interview G3].

By the end of the 1970s the city was notable as an industrial centre and remains so today. A town that had just over 20,000 inhabitants in 1960 had 120,000 in 1975 (Sánchez González, 1989: 82). Thus whilst a major part of Getafe's expansion was driven by migration to Madrid, the town's status as a significant employment centre within the Madrid region ensured that, from the start, it would be incorrect to consider it as a dormitory town for Madrid [Interview G11].

Since its being installed as the capital in 1581, in general terms Madrid's 'growth has neither been ordered nor agreed upon . . . so many books could be written on the plans made for a Madrid that was not built, a Madrid that was dreamed about' (Santos *et al.*, 2000: 321). More specifically, in contrast to the city of Madrid which has been subject to a succession of more or less elaborate plans since the mid-1800s (dal Cin *et al.*, 1994), central and regional government bodies have struggled to actively plan for growth in the broader metropolitan and regional space. The idea of a greater Madrid began to emerge within intellectual circles in the 1930s with talk of creating a 'nexus-city . . . a city without limits or frontiers' whereby transportation links would make the city the link between the fertile south and the industrial north of Spain (Santos *et al.*, 2000: 462 and 463 respectively). The growth of the city anticipated with such a role would be achieved through the incorporation of the whole region.

During the 1940s, ideas of aggrandising Madrid circulated alongside concerted efforts to industrialise the capital city and its region and the first signs of concern over an associated rise of a 'red belt' in the outer city (Neuman and Gavinha, 2005; Santos *et al.*, 2000: 528–535). Successive plans made provision to disperse industry and population from Madrid, where supply of land and premises were unable to meet the demands of neither industry nor migrant populations. In the immediate post-war period, urban development was relatively modest and channelled under the patronage of the Francoist regime (Castells, 1983: 218). Madrid's *Plan General* of 1946 was the first to look beyond the city of Madrid and envisage a radial-concentric pattern of transport infrastructure. This was a pattern reinforced by subsequent plans concerned with strategic planning of the metropolitan area such as the *Plan Metropolitano* (1963) and the *Esquema Director Regional* (1971). This was the radial-concentric framework upon which rapid growth occurred from the 1960s onwards at nodal points, notably to the south of Madrid (Ezquiaga *et al.*, 2000)

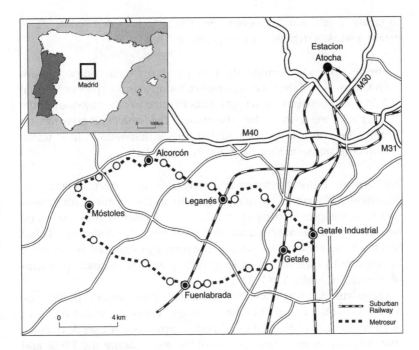

Figure 5.1 Map of Getafe

(Figure 5.1). Thus, by 1970, fully 64 per cent of the 3.1 million population of greater Madrid was concentrated in the 'new periphery' (Santos, *et al.*, 2000: 545). So, for example, today only 1.1 million people live within the M30 – Madrid's main ring road – compared to the 3.3 million beyond it (Verdú, 2003). It was also during the 1960s in the *Plan Metropolitano* that a green belt surrounding the city of Madrid was delineated.

Madrid city and parts of the metropolitan area of Madrid were planned directly by central government in the Francoist era by *La Comision de Planeamiento y Coordinacion del Area Metropolitana de Madrid* (COPLACO). In the 20 years from its establishment in 1963 to its dissolution in 1983, COPLACO produced numerous documents but no plans (Neuman, 1997: 82). Planning by central government dictat essentially left a strategic planning vacuum during this period within which the private sector was able to exert considerable influence (Neuman and Gavinha, 2005: 1007). Castells (1983) argues that an excessive centralisation of industrial activity overcrowded the major cities of Spain, attracting in thousands of workers from rural areas. This process was licensed by an authoritarian state where property developers worked closely with banks to create a vast building boom, and where lack of democratic rights meant a weak system of planning control. Thus

> the developers built hundreds of thousands of flats in compact groups in the middle of the Castillian plains, leaving empty spaces of several kilometers between clusters of blocks in order to raise the value of the land in between which they also owned. They only built housing – no amenities, no paved streets, no lighting, little sewerage, little water, and poor transportation . . . (Castells, 1983: 220)

With private-sector developers acquiring quite large tracts of land, one interviewee likened the pattern of growth in the southern municipalities to developers 'cutting up a cake' as and when demand for extra housing from migrant labour dictated [Interview G4]. Moreover, 'Madrid became a city of strangers, with . . . the recently urbanized areas unable to generate its own society as long as they were facing daily problems of survival' (Castells, 1983: 221).

The urban planning issues associated with these patterns of development were at the forefront of many citizens' demands, and at the heart of vocal and articulate neighbourhood groups – *asociaciones de vecinos* – that emerged in most large metropolitan areas during the 1960s and 1970s (Castells, 1983). Castells described these as some of 'the largest and most significant urban movements in Europe since 1945' (Castells, 1983: 215). Both in the historic centre of Madrid, as well as in the urban peripheries – in places such as Getafe – the improvement of the urban environment was at the forefront of political debate. Indeed, by the demise of the dictatorship the concerns of the municipalities in the southern periphery of Madrid and a recognition of the need to assist them had even belatedly forced their way into planning agendas of the City of Madrid and COPLACO (Neuman, 1997: 85).

In fact the southern cities grew in phases, with the closest to Madrid – Getafe, Leganés, Alcorcón – growing most rapidly in the 1960s and 1970s. Those furthest from Madrid – Móstoles, Fuenlabrada and Parla – grew at their most rapid during the 1980s. There is also a symmetry in the stabilisation and even decline in population of these same cities [Interview G4] (Ezquiaga *et al.*, 2000: 57). They are also experiencing the first signs of industrial migration – notably to smaller towns at the edge of the CAM territory where financial incentives and lower labour costs and rentals are available. Getafe in particular has already lost some of its Fordist manufacturing industry upon which it originally grew. Kelvinator has long since closed its factory in Getafe with the

Figure 5.2 New housing at the former Kelvinator site

site already having been redeveloped for housing (Figure 5.2), while John Deere's operations have been scaled back to marketing and repair and maintenance.[1] The poor condition of Getafe's industrial estates (Figure 5.7) and their poor connections to Getafe pose a current concern [Interview G5] as does the displacement of traditional manufacturing companies by logistics and service industries [Interview G6]. As a result of a combination of these factors, the metal working sector in Getafe is experiencing problems [Interview G6].

Since the formation of CAM, efforts at regional planning have varied in strength. Thus although 'the regional government develops a Strategic Regional Plan according to which the growth is controlled and directed . . . the reality is that this distribution of urban growth depends substantially on the potential of the existing transport systems and the capacity of each local council to apply political pressure. . . . each municipality is supreme in defining its urban growth, and most of them try to develop into their maximum possible size' (Maldonado, 2002: 364–365). CAM's limited formal powers over relatively autonomous planning powers of municipalities has meant that only periodically have influential personalities been able to exert a stronger regional planning influence [Interview G4]. The green belt established between Madrid city and the likes of the southern towns in the COPLACO era and maintained since then has been subject to on-going development pressure. In the light of this as one interviewee complained,

> But possibly this green belt will disappear... because the people build and central government cannot preserve this space. The municipalities are not interested in preserving this space because they need money to develop the city and the main money comes from the building of housing. [Interview G4]

The desirability of any merging of Madrid city and southern towns is, from the perspective of the latter, highly questionable. 'Madrid has reached a type of crest in terms of growth... nobody wants Madrid to become this monster city – there are questions such as that citizens are more distant from the government' [Interview G1]. Nevertheless, the loss of green belt land and the continued expansion of the southern cities recently prompted one Barcelona newspaper to run a feature article exclaiming that 'Madrid could become a kind of Los Angeles'. If the rates of growth being experienced in Madrid in the late 1990s and early 2000s were to continue, by 2035 settlements such as Getafe would be merely part of one continuous urban fabric covering the whole of the CAM territory. In all of this there is a sense of history repeating itself. According to urban planner Miguel Colmenares, 'Madrid and its metropolitan area have experienced in a decade in terms of speculation what took place in Europe over fifty years.' Some sense of the lack of strategic planning for accommodating growth in the Madrid city-region is implied in the fact that the 700,000 square meters of land being released for development in 2003 in the CAM territory is triple that in 2001 (Rodriguez, 2003).

As a result of this largely unregulated expansion, then, 'in the 1980s, they are very populated cities but without any kind of facilities. So the planning of these cities in the 1980s tried to solve all these kinds of problems. To give schools, to give a university, to give a music school... to give a cultural space, to give a sports space... to try to finish the city' [Interview G4]. The problems created by unregulated expansion also went beyond the local under-provision of amenities and services with significant problems of congestion being caused by the growth of 'transversal' commuting among the southern municipalities [Interview G4]. Indeed, this is precisely why the Metrosur project was conceived in order to not only ease traffic congestion but also create a degree of economic cohesion.

Successive Getafe local plans for 1969, 1979, 1986, 1995 and 2003 have been illustrative of the sort of competition for growth that has existed among the towns and cities in the south of the Madrid metropolitan region. Whilst the 1969 plan was the first to reserve land for much needed facilities and amenities in Getafe, the subsequent 1979 plan was excessively expansionistic. The current 2003 plan still makes provision for further expansion in the form of 10,000 new homes [Interview G7].

Thus, the planning arrangements referred to here actually continued trends of localised expansion at each of these settlements despite the formation of the CAM to coordinate planning at a regional level, since development has provided the means for securing an improved fiscal base for the municipalities. Here the southern municipalities are part of more diffused patterns of urban development that have seen a rebalancing of the economy and population in the CAM region. 'The traditional relationships of dependence between the centre and the periphery appear to have altered substantially owing to the appearance of new urban centres in areas previously considered peripheral' (Ezquiaga *et al.*, 2000: 55).

Different opinions are evident regarding the degree of co-operation and competition existing between the southern municipalities. It is curious, for instance, that there is relatively little competition among the southern municipalities for private investments in the streamlined and regionally coordinated realm of investment promotion in the CAM territory [Interview G2]. However, 'with the arrival of democracy we see a certain degree of competition starting up between the towns – all the mayors want their town to be the best – so everyone tried to improve things' [Interview G1]. By the 1980s the lack of strategic planning in the Madrid metropolitan region had left a situation among the Southern municipalities neatly summarised by Neuman (1997).

Its seven towns were 'municipalities adrift'... The South began to take on a weight of its own and develop its own approach to urban development and management. Lacking higher-level support, each town went its own way. The municipalities began to compete with each other for new growth in a classic 'fiscal zoning' fashion. The towns were 'dominated by the logic of the real estate market'. (Neuman, 1997: 86)

As the result of the weaknesses of strategic planning at the regional level, even within the higher-level context provided by the CAM, there remains significant competition for regional and central government resources and an aggressive pursuit of an expanded population and local tax base by each of the municipalities. The persistence of this competitive expansionism was described by one interviewee:

> Well firstly they are independent towns – each has its own plans, its own idiosyncracies – the only thing that unites them is the regional planning from the Comunidad de Madrid... Between the municipalities there is nothing more than cordial relations. There is no common strategy – it is the Comunidad de Madrid which decides the common strategy. Really it is a collection of small kingdoms. [Interview G7]

As we explore below, among these competing local kingdoms, Getafe has been one of the more vigorous municipalities to exploit this space for action.

5.3   Getafe and the 'Gran Sur': Political mobilisation within the metropolitan space

The political settlement that followed the collapse of the Francoist dictatorship was characterised by a strong degree of mobilisation at the local level. The improvement of the environment was at the forefront of the political debate promoted by grass-roots movements, and the role of the newly democratic state – especially at the municipal level – was seen to be fundamental in achieving this.

The protests in southern suburbs such as Getafe, Leganés and Móstoles, which drew attention to this lack of facilities, were among the most militant displays of opposition to the Francoist regime and they conferred a lasting political legacy. First, the very success of the grass-roots movements brought about their co-optation into local political structures with leading activists going on to hold prominent positions in municipalities throughout the region (Maldonado, 2002: 366). Second, the strength of some of the southern municipalities rests not so much in their administrative or statutory autonomy but in terms of strength of political backing. The cities of Getafe, Fuenlabrada and Leganés have enjoyed the strongest political backing, having a consistent record of returning a socialist mayor. Alcorcón and Móstoles have had more changeable political complexions [Interview G4]. Third, as Castells (1983: 262) notes, perhaps the most important contribution of grass-roots movements was to add a 'social' conception of the

Spanish city to sit alongside the bureaucratic and capitalistic conceptions hitherto promoted exclusively under the Franco regime.

With their common origins in the industrialisation of Spain there
has come to be a strong shared identity among the municipalities of
the south of Madrid. As one interviewee described, 'there is a sense
of identity... People in the south recognise that they are from the
south of Madrid and they are in the industrial heartland' [Interview
G2]. Getafe as the first among the municipalities to expand and the
most closely associated with Fordist manufacturing industry has a
population with perhaps a special character which in turn has trans-
lated into a strong base of local political power. As one interviewee
described,

> The people of Getafe are workers in industry. They have worked in
> industry since the 1940s. They had to fight against the dictator and
> probably they are better connected with trade unions.... And I think
> that probably could be the reason that they are very engaged all the
> time and are discussing all the time with the central government.
> [Interview G4]

A pertinent example here would be the Mayor and the municipality's
intervention in the construction of new toll motorways by the central
government ministry of works. These new motorways have exercised all
of the southern municipalities, since there are concerns that they further
penalise the south in relation to the wealthier north of Madrid, and will
physically divide communities and create additional financial burdens
on businesses and workers. The municipality of Getafe led by the Mayor
have been perhaps the most prominent among southern municipalities
in opposing the works and seeking to negotiate benefits or ameliorate
impacts [Interview G4].

During the 1980s, the social democrat (PSOE)–controlled CAM
attempted to combine a territorial planning strategy with an electoral programme, which sought to redistribute wealth from the north
and north-west of the city, identifying four major territorial lines of
action of which the South was one of the more important. The major
municipalities of which the South consisted – Móstoles, Leganés, Getafe,
Fuenlabrada, Parla and Alcorcón – were seen as fragmented, in need of
coordinated governance (Heitkamp, 2000; Neuman, 1997). Felix Arias,
director of the *Oficina de Planeamiento Territorial* (regional planning
office), identified the 'Ciudad del Sur' (the city of the South) as a means

of endowing identity to the fragmented southern municipalities, and creating a kind of second city to sit alongside the Madrid core:

The city of the South, a city of more than 800,000 inhabitants at present, thus raises itself as a city with an airport and university, with parks, sports and leisure facilities, with important urban and commercial centres and with the creation of specialised areas for economic development that currently can be decentralised in the Madrid metropolis, but can't find suitable land in the metropolitan South, such as business parks, transport nodes etc. (Arias, 1991 [speaking in 1988], p. 426)

The concept of a whole new city was given credence by the excellent transport links in the south, with the convergence of four major motorways, a mainline train service to the south of Spain, and a number of established urban centres. It was also given necessity by the growth in unemployment in Spain during the 1980s, particularly in the industrial areas to the south of Madrid.

Beyond the political base, Getafe has also managed to distinguish itself not merely among the southern municipalities but also within the CAM area in terms of important institutional innovations and in what Neuman terms 'technopolitical' patterns of working to promote itself. As a representative of Getafe Iniciativas (GISA), the privatised economic development arm of the municipality, described, 'this is a council that is very advanced in the field of local economic development, for fifteen years it has had an instrument separate from the council and the other councils don't have this' [Interview G8]. Moreover, despite the common political leanings of the southern municipalities, the Mayor and senior council officers have pursued patterns of pragmatic technopolitical negotiation that have been successful in delivering local economic development. As the same interviewee identified, 'the rest of the municipalities of the South, they're left run councils and nonetheless none of them have the local development approach conceived in Getafe' [Interview G8].

5.4   The making of the 'capital of the south'

Post-1979, the development of local capacity for governance in Getafe has taken place against, and drawn strength from, a threefold dynamic of politico-economic restructuring. First, despite some restructuring of the local industrial base associated with the restructuring of the

Spanish Fordist economic model (Holman, 1996), Getafe's status as one of the most important sites of industrial development within the Madrid metropolitan area remains. Second, the re-establishment of democratic municipalities was consolidated after the local elections of 1979. The Ayuntamiento de Getafe was given responsibility for local planning, which includes considerable powers for co-financing urban projects. Third, there has been the creation of a regional tier in Spanish governance, in this case the CAM, with responsibilities now including strategic planning, education and strategic economic development. Led by the *alcalde* (mayor), the social democrat Pedro Castro, Getafe has been able to alter its own political weight through negotiation with the regional presidency. As such, it may not be implausible to see Getafe as an entrepreneurial local state, which is defined by its location both on the southern edge of Madrid and within a wider strategic metropolitan context, as we now explore.

Our earlier discussion of the boundary-transcending activities of local institutions and agents (Cox, 1998) leads us to focus on Getafe in terms not only of its formal administrative powers or its economic strength but more so on specific aspects of the geopolitical power available to the municipality and its local politicians. The grass-roots political movements appear to be part of Getafe's enduring geopolitical capacities. First, Getafe has a place or civic identity capable of mobilisation by a local state fraction. Second, its real, virtual and symbolic territorial location within the Madrid political space has also been exploited recently.

The characteristic features of southern European urban politics have been described as 'an overlapping of authorities and regulations, competition for power and resources, and government officials acting with increased freedom and capability of exploiting the various representative and governmental institutions' (Delgado *et al.*, 1998: 238). Within this context the political capabilities of local mayors is of key importance. However, recently the role of the mayor has changed somewhat from traditional clientelism along party-political lines to horizontal lines with construction of political leadership and territorial mobilisation. As Barroz and John (2004: 118) have emphasised recently, in southern Europe 'leaders must find and build their legitimacy locally to the point where there is often little future for them outside their communities'. Getafe's Mayor, Pedro Castro, has proved himself adept in operating on both fronts mobilising territorial identity to exploit regional and central governmental institutions and in the process sustaining his own political base – but to the point that he is perhaps one of only two notoriously successful mayors among the municipalities of the CAM region [Interview G4]. What is less clear is whether the case of Getafe is an example of 'a greater collaboration between the public and the private sector in the framework of defining and setting up public projects' said to be apparent in Spain (Geneys *et al.*, 2004: 196).

Getafe council has promoted itself as 'the capital of the south', the lynchpin of the southern Madrid working-class towns, which has an underlying geopolitical rationale. Within this evolving 'metropolitan space', the Mayor of Getafe, Pedro Castro, has played a key role in aggrandising Getafe not merely as part of the city of the South but as 'capital of the South', thus enhancing both his and his town's predominance in the new city-region. As one interviewee described, 'the Mayor of Getafe now is a very clever person... From a political point of view. So he tries to obtain for his municipality things that the other municipalities don't have' [Interview G4]. There are several features of this that are worth identifying.

First, as we have seen earlier, Pedro Castro can call on historically strong and consistent socialist party-political support locally but also on his own popularity which he has sought to renew when capturing flagship developments for Getafe. Second, his style of working is populist in that it is geared to circumventing administrative bottlenecks.

> He [Pedro Castro] is a very special person. The other people try to work at the same level as the regional government. We try to sign an agreement. We try to approve things within the law. Pedro Castro is not the same. He wants to do these things and he thinks of the way to do these things without the law and without planning... [Interview G4].

Here Castro appears to have been a skilled operator in the part technocratic, part political process of imagining the Gran Sur. As Neuman describes it, 'the technopolitical process used to reach agreement between the regional government and the southern municipalities was 'ad hoc, outside statutory planning procedures' (Neuman, 1997: 90–91). Third, what differentiates Pedro Castro from the other mayors in the red belt of the south has been his ability to work pragmatically with politicians of different persuasions at both central and regional level. The mayors of neighbouring municipalities, for example, have been far less successful in their dealings with regional government even when they have come from the same political party [Interview G4].

Mayor Pedro Castro has been in power for over 20 years in Getafe, during which time he has managed to fashion the municipality as the capital of the south. As he has recently outlined, 'it has fallen to our city to be the motor of the South [of Madrid]. Getafe is the centre of economic development it is the university and cultural centre' (Getafe.net, 1999). A mainstay of each of the southern municipalities has been the generous allocation of land for housing development but Getafe is perhaps distinguished by the aggressive and pragmatic pursuit of flagship economic, social and educational and leisure developments by the municipal leaders. In the economic sphere alone one interviewee distinguished Getafe within the CAM territory, 'with Getafe, unlike many other places, there is a programme to use land for industrial purposes – industrial estates and so on – and this of course draws business here' [Interview G9].

> because of the Mayor, Getafe could be seen as the pole of reference of the southern area of Madrid and it is in fact called the capital of the south. This irritates the other mayors and their residents but it is the reality – Getafe leads progress in the southern area. The most advanced initiatives come from Getafe and it leads as it were that special understanding with the Comunidad de Madrid. That is the reality. [Interview G7]

Indeed, the continued development of Getafe in particular and of the south of Madrid as a whole has produced a dynamic that is striking in national terms. As one interviewee vividly describes,

> traditionally the south of Madrid has been a depressed area – a workers' area. When I came here I thought I would find a town but I found a city – Getafe is bigger than many provincial capitals in Spain. It's bigger than Salamanca, and I was also surprised by the amount of business going on here.... when you speak about the town itself... it has undergone a great transformation compared to what it was. And the south has become the engine of the Comunidad. [Interview G9]

In the light of this, Ruiz-Gallardón, president of the CAM, has recently referred to the south as Spain's second or third city [Interview G4].

There are several examples of the sort of flagship developments that have been captured for Getafe and in which the mayor and municipality officers have demonstrated their technopolitical skills to which we now turn.

One element to this rethinking of the municipality's role within the *Gran Sur* has been the 'integration of the different districts of the city through neighbourhood regeneration aimed at improving their quality of life and endowing them with an identity that induces the residents' belonging' (Castro, 1999: 12). The sinking of a suburban railway line that has divided Getafe provides one important example of attempts to unify the municipality and create a sense of place (Figure 5.3). Getafe was the first among the other southern municipalities to seek do this at a time when central and regional government funds were available. Here again Getafe has managed to secure benefits that other southern towns will struggle to obtain. This is also another instance of the nego-tiating skills of and political backing enjoyed by the Mayor and the municipality more generally seen earlier in the case of the new toll motorway.

> With the Comunidad de Madrid, there's a good relationship... this Mayor, this Mayor is very belligerent, so he gets a lot of things out of the Comunidad. With the national level it's very bad. In fact, there was a very important war over the financing of the works for the tunnelling of the main train line [Interview 8].

Figure 5.3   The burying of the suburban railway line

This particular 'war' was over the fulfilment of an agreement by the central government to co-fund the sinking of the railway line after a change from socialist (PSOE) to conservative (Partido Popular) political control.

The commissioning of Norman Foster's architectural practice to produce a master plan for a 43-hectare site at the southern end of the municipality is, perhaps, the most notable instance through which Getafe's identity has been promoted through regeneration. Norman Foster was originally commissioned by the CAM to plan for the development of a major site made possible with the release of land adjacent to Getafe's military airbase. Foster's 'Aeropolis' plan – based around exploiting synergies with Getafe's oldest employer, CASA, involving manufacturing and logistics – was regarded at local level in Getafe as utopian and unworkable. The strategic importance of this land could hardly be understated since, as one interviewee noted, 'this is the largest area of land in region. It is one of the largest developments in Europe and without doubt the largest in European capital city-regions' [Interview G2]. It highlights the manner in which major peripheral urban land uses such as 'airspaces' are subject to significant transformation over time (Pascoe, 2001).

More to the point, the progress of this plan also highlights yet again the political capabilities of Mayor Pedro Castro. He lent his support to the plan despite a degree of controversy given that it was parachuted in by CAM on the basis of Getafe's accessibility for such a strategic development. As a result, the contents of the plan were opened up to a broader discussion and examination and perhaps in the process elements of the plan have been diluted and usurped. Pedro Castro contrived to convene his own press conference to unveil the scheme one day prior to CAM's own press conference. As one interviewee observed,

> Foster made a very beautiful plan. This project was covered in the press . . . and they did nothing of the Foster plan. They develop this land but with other projects. So the importance of Foster is the glamour, the image of the city. Foster, the great architect, makes a plan for Getafe [Interview G4].

In this instance, the municipality has already pushed ahead with developing part of the site for logistics park leaving sites potentially more difficult to market to be developed by the CAM.

The other southern municipalities have tried to follow suit in these efforts but have so far been unsuccessful. 'Getafe is like a symbol for the other municipalities. And if Getafe tries to obtain anything then they also want to obtain these things' [Interview G4]. A lack of funds has prevented the burying of rail lines in the other southern municipalities while the location of Carlos III University in Getafe is certainly a coup in the light of its size and academic standing, which is unlikely to be repeated in any of the other southern towns [Interview G4].

The defeat of the social democrat PSOE by the conservative Partido Popular in the 1991 regional elections has seen the right control the CAM for more than a decade, under the presidency of Alberto Ruiz-Gallardón. However, the CAM president was considered very much a pragmatic, centrist politician who was dedicated to strong regional intervention in modernising the Madrid metropolis as a competitive city-region. Here, Getafe plays a key role as an industrial location – the municipality is to hold 80 per cent of the CAM's new industrial space between 2000 and 2002 (Fernandez, 2000).

Getafe's Mayor, Pedro Castro, and the former CAM president reached a controversial accord over the construction of the latter's pet project, the Metrosur, an extension to the Madrid metro which would link in all the major municipalities to the south of the core city (see Figures 5.1 and 5.4). Just prior to the opening of the Metrosur, one interviewee described how 'the southern municipalities aren't in agreement

Figure 5.4   Getafe Centro Metro station and the John Deere factory

with this, but Getafe is... although it's been modified a little – the Metrosur route... Getafe has always been for the metro because it thinks the connection with the other municipalities is the most important thing for reasons of economics' [Interview G8]. Undoubtedly, for Ruiz-Gallardón, the Metrosur would allow a dilution of the southern red belt through increased commuting possibilities, as well as enhancing mobility within the region. Yet, unlike the other southern mayors, who saw Ruiz-Gallardón's announcement as being either an electoral false promise or an attempt to 'gentrify' the south, Getafe's Mayor seized the opportunity and offered public support in return for extra stations within Getafe's neighbourhoods. The Metrosur represents 20 per cent of the total Madrid metro network but serves only 5 per cent of its users, prompting one newspaper article to note how 'the southern municipalities, historic fiefdoms of the left, were to be seduced to vote for the Popular Party by the charm of rails' (Verdú, 2003).

It is a testimony to Pedro Castro's technopolitical skills that there was, for example, little concern in the business community over the impacts of any change in political control of CAM.

Getafe has had the PSOE for some 24 years, whilst the Partido Popular in Madrid has been in power for eight and they have followed essentially the same strategy. Pedro Castro has many friends in the Comunidad from the Partido Popular – of course social policy is going to be different – but business policy is the same. We recently had elections for the Comunidad and there was no great concern on the part of business over who would win. [Interview G9]

In 2003 the socialists gained control in the elections in June 2003 and seemed able to form a narrow majority government with the post-communist Izquierda Unida (United Left) – however, in one of Spain's more bizarre political twists, two PSOE deputies did not attend the debate to elect the new president, instead absenting themselves in a hotel that later emerged had been paid for by property developers. New elections were held again in October 2003 producing a narrow Partido Popular (PP) victory. In the intervening months the deputies were expelled from the PSOE, and there was much speculation concerning their relationship with property developers who were believed to be close to the PP. This scandal over property speculation, business and its relationship to political parties has also touched Getafe (Figure 5.5), where the opposition PP called for high ranking members of Getafe council – including Pedro Castro – to be investigated over apparent

Figure 5.5 PSOE Getafe – the clan of all the thousand million

irregularities (for example, matters related to the redevelopment of the former Kelvinator site). Following 22 days of investigation through a local commission, the allegations were found to be unproven (Hidalgo, 2003).

The Metrosur was conceived by CAM as a strategic planning objective in order to integrate the southern towns and alleviate transversal traffic flows among them. However, despite these strategic objectives and despite Getafe gaining extra metro stations in the process of Mayor Pedro Castro's highly effective manoeuvring in the metropolitan political space, the Metrosur route in Getafe crucially did not take in the municipality's important industrial estates. This story reflects the wider pattern of business interests within local governance, a consideration of which we now turn to in the following section.

5.5 Party politics, personality and the politics of place-making

To begin with, we can say that there are important similarities with the pattern of business interest representation – or rather disengagement at the local level – found in our other southern European case-study post-suburban municipality. Here the lobbying role of collective business organisations is underdeveloped in comparison to their service providing and social network roles. Whilst collective business groups proliferate, the link between business interests and local mobilisation is severed by the manner in which separation of residence from work place takes effect. As one interviewee observed, 'Spanish businessmen are quite different from say Germany or the UK and are very individualistic. Also very few who have businesses in Getafe live in Getafe, and therefore they don't feel identified with Getafe' [Interview G1].

However, in contrast to the case of Kifissia, the other southern post-suburban municipality considered in this study, local business interests, whether defined at the municipal, sub-metropolitan (i.e. southern municipalities), metropolitan and regional level, have become quite organised, following a pattern that appears nationally (Figure 5.6). As one interviewee described it concerning the organisation of business interests in the CAM region in general, 'there are associations for absolutely everything. In the centre of Madrid every single street has an association. . . . There are associations for anything you care to mention but they don't necessarily serve any purpose' [Interview G2].

A number of local branches of business representative bodies have been established reflecting in part the proliferation of such associations nationally and partly the size and maturity of Getafe as an industrial centre. These included local branches of national organisations such as the Cámara de Comercio and particular sectors such as the metal workers. There are also a number of fora that have emerged as conduits of business views into council matters. A Commercial Sector Board meets

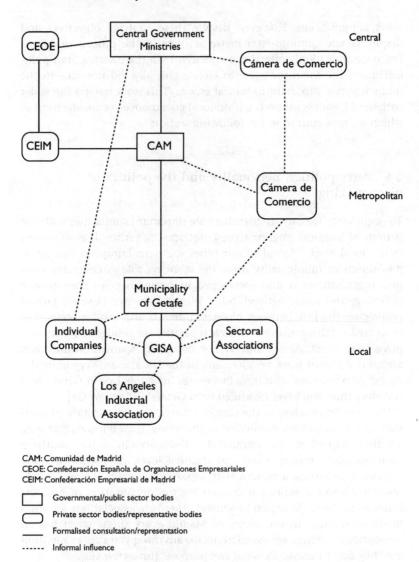

CAM: Comunidad de Madrid
CEOE: Confederación Española de Organizaciones Empresariales
CEIM: Confederación Empresarial de Madrid

☐ Governmental/public sector bodies
⬭ Private sector bodies/representative bodies
— Formalised consultation/representation
----- Informal influence

Figure 5.6 Map of business interest representation in Getafe

quarterly with local politicians and officers from the municipality in attendance [Interview G11].

The formation of GISA, the private economic development arm of the municipality, is itself, as we noted earlier, relatively unique within the

CAM territory. This is testimony to the economic development orientation of the municipality, but some further evidence of the amenability of
the municipality to private-sector interests is highlighted by the degree
of rationalisation or streamlining of these bodies. In an initiative led by
GISA, the main business representative organisations have been housed
along with key public-sector service-providing organisations, such as
laboratories of the Fundación de Innovación, in single shared premises
on a major industrial estate and with good access to GISA – the private-
sector economic development arm of the municipality.

In general, these local business representative bodies did not
appear to have strong views regarding local issues in policy-making
reflecting a national pattern. The exception would be representatives
from the metalworking sector, long Getafe's key sector, who voiced
concerns about the lack of local institutional support for the industry
[Interview G6].

Yet, the growing role and influence of business in urban development
in Getafe owes a great deal to, and is channelled through, the pragmatic
mayoral politics of Pedro Castro. Mayor Pedro Castro is at the centre
of Getafe's urban entrepreneurialism, which in broad terms has seen a
growing dialogue with business and a pro-business outlook in municipal agendas. Yet, the importance of the mayor for channelling these
interests is vividly highlighted in the signal failure of the business voice
to be heard regarding the route of the new Metrosur metro line through
Getafe, even when GISA, the municipality's own commercial arm was
representing local business views. As an interviewee recounted,

> The route was decided by the politicians. In both cases in Getafe
> and Fuenlabrada, there were two examples where industry really did
> try to influence the route but their views were overlooked.... Here
> we have representatives of the most populous area outside the core
> nucleus of the capital city and they have no power to get their bus
> stops or their train stations put in. And they tried, they really did, on
> behalf of business and with the businesses... [Interview G2]

The clear implication is that this important infrastructure development was obtained and bargained over entirely within the political
and administrative machinery of local and regional government with
little or no reference to the views of significant organised private-sector
interests.

The significant deficit of links between the industrial zones and
housing areas in Getafe as a result of business failure to influence the

route of the Metrosur is presently a source of considerable disquiet among the business community [Interview G9] with companies on the industrial estates financing transport connections for workers. The inadequacy of such transport links comes on the back of long-standing concerns regarding the general condition of Getafe's original 1940s- and 1950s-built industrial estates [Interviews G5, G6, G12] (Figure 5.7).

Here, after a period of state-sponsored but private-sector-constructed rapid urban growth and corresponding growth of socialist politics at the municipal level, a new period of municipal engagement with the private sector is just apparent, reflecting a more general trend among Spanish municipal leadership (Barroz and John, 2004). In the case of Getafe, interviewees identified an influential role for business inputs into decision-making. However, 'administration – local and regional – are the most influential. They make the final decisions whether it's to be industry or housing. It's the institutions that decide if we are going to invest in such an area or not' [Interview G1]. In this case of recent developments in Getafe, then, local municipal and regional government policy-making has yet to incorporate business views to any significant extent and, on the very biggest of issues, has proved quite impervious to business influence.

Figure 5.7   Los Angeles industrial estate, Getafe

5.6 Conclusion

Getafe began life as a post-suburban space imposed by the central state, but in this instance a sense of place was constructed out of grass-roots political concerns with the attendant social and welfare problems of this imposition. These grass-roots movements conferred a lasting local political capacity that has been mobilised in the mayoral politics of Pedro Castro. In contrast to the case of Croydon which we review in Chapter 8, it is Mayor Castro's political manoeuvring, rather than the mobilisation of local administrative capacities *per se*, that has raised the profile of Getafe as a distinct place with distinct concerns within the wider metropolitan setting. Mayor Castro has been able to mobilise the grass-roots political legacy to enlarge his own spaces of engagement (Cox, 1998) within the metropolitan and national government spheres and bargain for additional investment to fashion a fuller sense of place identity in this state-created post-suburban space.

In appearance, Getafe is quite unlike a US-style edge city. However, speculative private-sector development was central to its rapid growth as a dormitory suburb even though this role has been complemented by additional employment, leisure and educational land uses. Whilst business interests are highly organised at the local, metropolitan and national level within Spain, there is little evidence that these interests have played any significant role in driving development agendas in post-suburban Getafe. Instead, the development of Getafe and its emergence as the 'capital of the south' has embodied a successful balancing of continuing pressures of private-sector-led development, on the one hand, with collective consumption expenditures fought for through party political alliances and prized from government structures and funding streams on the other.

6
Noisy-le-Grand: Grand State Vision or Noise about Nowhere?

> Where did they all go? To the outskirts. To the suburbs. Paris had become a business hypermarket and a cultural Disneyland.... And if Paris had emptied, if it was no more than a ghost town, didn't that mean the true centre was now 'all round'?
>
> François Maspero, *Roissy Express*

6.1 Introduction

Local government and the public sector more generally play an important role in the development of each of our case-study edge municipalities, in a way they do not in the North American setting. However, nowhere is the influence of non-local State institutions and constructions more apparent than in Noisy-le-Grand whose edge identity, or lack thereof, has been produced by its entanglement in a complicated and overlapping set of administrative arrangements.

Noisy-le-Grand exists as a *commune* located 12 km east of Paris (Figure 6.1) and is a settlement that dates back at least as far as the invasion of Gaul by Julius Caesar (58–52 BC). The first settlement may even date back as far as the iron age (800–450 BC). Noisy-le-Grand remained a village until the beginning of the twentieth century, but started expanding in the inter-war period, increasing from 2200 inhabitants in 1921 to over 10,000 in 1954 (Ville de Noisy-le-Grand, 2003: 8–9). It was the main centre of employment and population growth in the east of Paris in the 1950s and 1960s, having reached a population of 25,800 by 1968. However, the main development of the town has been from the 1970s onwards with the implementation of regional development plans aimed at controlling the growth of Paris resulting in Noisy-le-Grand's incorporation into the planned new town of Marne-la-Vallée. As a result,

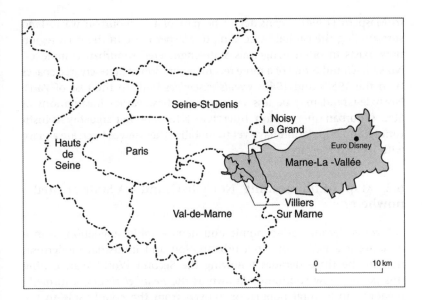

Figure 6.1 Map of Noisy-le-Grand

Noisy-le-Grand's population doubled from 26,765 in 1975 to 52,408 by the end of the 1980s.

In the 1990s, the rate of growth of the population slowed down. At the end of the 1990s the population had risen to just over 58,000 (INSEE, 1999). Along with the increase in population has come an increase in employment in the town, both being linked to the development of the new town and concentrated in the new areas of development.[1] At the end of 2002, two-thirds of employment was concentrated in Mont d'Est and Pavé Neuf, the heart of the new tertiary districts developed since the creation of the new town. When the Richardets industrial zone, created in 1974, and Champy, home to several higher education establishments, are added to this, the newly developed areas of the town account for over 80 per cent of employment in Noisy-le-Grand. Indeed, the dominance of the Mont d'Est business and commercial centre built in the 1970s has grown from 38.3 per cent of employment in the town in 1988 to 56 per cent at the end of 2002 (Balaquer *et al.*, 1996: 15; DDEE, 2003a: 6).

The development of Noisy-le-Grand can be seen as a product of traditional French statist approaches to regional planning and economic development. Reflecting a stronger and long-standing tendency for undesirable populations and land uses to be expelled to the edges of European cities, Noisy-le-Grand is part of what Parisians – according to Maspero (1994) – consider to be part of the suburban waste land surrounding the capital. According to Maspero, life in the Paris region now exists in often dangerous State-engineered suburban complexes. Noisy-le-Grand is one of a more recent generation of State-created spaces than the 1950s and 1960s *grands ensembles* built to the east of Paris. Noisy-le-Grand may be less violent than these earlier incarnations of State suburban intervention [Interviews N13, N14] but arguably its institutions and population have yet to mobilise themselves to create a place with its own identity.

6.2   Maps of no meaning: Noisy-le-Grand – A State-created nowhere?

Following decades of economic and demographic stagnation, France underwent a period of rapid urbanisation and industrial modernisation in the three decades following the Second World War. During this period, now commonly known as the *trente glorieuses*, a formerly predominantly rural population moved from the countryside to the growing urban centres of employment. At the forefront of these trends was the capital city, Paris, which consequently witnessed large-scale, cheap housing developments, or *grands ensembles*, around its edges to accommodate the rapid influx of new workers. By the 1960s, several problems had become apparent with this model of State-directed industrial modernisation. First, Paris was capturing French economic and population growth to the detriment of balanced growth in the rest of the country, prompting Jean-François Gravier (1947) to speak critically, as early as 1947, of *Paris et le désert français*. Second, it was evident that Paris was suffering from major problems connected with congestion and poor housing. Even on the outskirts of the city, the new *grands ensembles*, cruelly lacking in local amenities and infrastructures, soon became associated with the kind of boredom, alienation and social problems superbly captured in Christianne Rochefort's 1961 novel *Les petits enfants du siècle*. Indeed, one site of such housing developments, Sarcelles in the north of Paris, gave rise to the term 'sarcellitis' to describe the social and psychological effects of life in such places.

In order to overcome the dysfunctional aspects of rapid industrial growth, the French State – which played a predominant role in economic development within a *dirigiste* framework – attempted to halt the growth of Paris by moving industries out of the capital and blocking new developments within it. As part of this strategy for achieving a more balanced urban growth by creating jobs, housing, cultural facilities and services outside of the main urban agglomerations of the country, nine new towns were planned near major cities, five of them in the Paris region. One of these, Marne-la-Vallée to the east of Paris, was programmed in the 1965 *Schéma directeur de la Région Ile-de-France* (SDRIF) in order to overcome imbalances caused by the historical absence of industry and development to the east of the capital, and includes the pre-existing rural *commune* of Noisy-le-Grand. It was not until 1971, however, that the creation of Marne-la-Vallée was announced, and not until the following year that the Epamarne was created to manage and implement policies for the development of the new town (Balaquer *et al.*, 1996: 7).

Marne-la-Vallée was not planned as a new town in the Howardian tradition. Rather than assembling residents in one locality in a more or less concentrated manner, the new town of Marne-la-Vallée was to be a series of small-scale settlements based upon existing *communes* and around transport connections (road and rail). Following the Marne river for 25 km, to a depth of 4 km, from Noisy-le-Grand in the east to Saint-Germain-sur-Morin in the west, these were to be separated by extensive parks, lakes and woodland and were to provide a mixture of collective and individual housing, social amenities and local employment in a rationally planned way. The essence of the concept, captured in the slogan of the new town, *la campagne dans la ville* (the countryside in the town), can be contrasted with the previous phase of high-density development around Paris.

In 1969, before development work began on the new town, it numbered 87,000 inhabitants within its perimeter (Epamarne/Epafrance, 1999). Thereafter growth was rapid throughout the 1970s and 1980s. Although this slowed down in the 1990s, the 1999 census showed a total of 246,652 inhabitants, and Epamarne estimated that this had grown to 268,200 by January 2003 (Epamarne/Epafrance, 2003: 10). This growth is quite different in both form and process to urban sprawl in North America, based as it is upon the *intentions* of rational State planning and mixed-use communities linked by efficient public transport systems. Yet there are some minor parallels with North America in terms of the symbolism of new town development.

Pioneers of the avant garde

As a geographic and functional entity of just under 60,000 inhabitants providing 25,400 jobs, Noisy-le-Grand obviously cannot be termed a city. Nevertheless it has some characteristics in common with Garreau's (1991) concept of an edge city.

Garreau (1991) suggests that edge cities represent frontier regions. While it would be a vast over-exaggeration to suggest Noisy-le-Grand is a frontier town in the North American sense, the rapid expansion of the area from a small rural town to a tertiary centre from the mid-1970s onwards has conferred a pioneer status to the town. First, populations were moved into the town from 1974 onwards, ahead of the development of local amenities and transport infrastructures – the A4 motorway connecting the town to Paris was opened in 1976 and the RER rail connexion reached Noisy-le-Grand only in 1977. Indeed, Dieudonné (1992: 27) shows that early settlers in the area considered themselves to be 'pioneers', attempting to create a sense of place in a previously uninhabited space.

In the early days, a delay in putting essential infrastructure in place slowed the growth of Marne-la-Vallée. In addition, the deterioration of the economic climate, with recession hitting France in the late 1970s, also slowed the growth of the town. Despite the delays these obstacles caused in the building programme, by the mid-1980s, 400,000 m² of office space were available, and the Mont d'Est district of Noisy-le-Grand accounted for two-thirds of employment in Marne-la-Vallée (Balaquer *et al.*, 1996: 8–9). The jobs created were mainly in the tertiary sector. In 1994, a new SDRIF confirmed the vocation of Marne-la-Vallée as a counterweight to the development of the west of Paris, and the economic role of the first sector (covering Noisy-le-Grand and the neighbouring *commune* of Villiers-sur-Marne) within this through its role as a centre of tertiary employment. Indeed, the SDRIF saw the first sector as an urban centre of European standing and aims at a doubling of economic activity and office space by 2020 (Noisy-le-Grand and Villiers-sur-Marne, 1999: 7).

Second, Noisy-le-Grand has tried to play on this sense of being on a frontier. In the early years of the town's development, attempts were made to fashion a local identity based on the idea that it was an 'experimental' town. Thus, early large-scale building projects included not only the (then) ultra-modern business district of Mont d'Est but also an attempt to combine commercial, consumption and residential functions in one space by constructing offices and flats on the roof of the Arcades shopping complex. Indeed, experimental architecture figures prominently in Noisy-le-Grand and distinguishes it from the suburban *grands ensembles* housing developments of the 1950s and 1960s elsewhere (Noin and White, 1997). The most notable of these were the experimental residential complexes of 'Les camemberts' in Pavé Neuf and 'Le Théâtre' in Mont d'Est.

Even today, Noisy-le-Grand continues, albeit with less vigour, to promote itself as a town at the cutting edge of technological developments – as a sort of 'pilot town'. Thus in 1998, the municipality was able to proudly announce that France Télécom had chosen Noisy-le-Grand as the site for the first European experiment in Asymetric Digital Subscriber

Line (ASDL) Internet technology (DDEE, 1998). In the same year, the town was also used to experiment a combined travel, telephone and payment card by the Parisian transport provider RATP (CVE, 1999).

Such architectural and technological innovations have been trans-lated into the town's office building programmes, including the ambi-tious 270,000 m² Horizon Paris office development (*Noisy magazine*, 2001a). Advanced eco-friendly and fibre optic technologies are vaunted as prime selling points alongside the low price of land and office space compared to the nearby capital city, and feed into municipal strategies for promoting Noisy-le-Grand as the 'the major tertiary centre in the eastern Paris region' (Ville de Noisy-le-Grand, 2004).

Noisy-le-Grand was intended to counterbalance the development of tertiary industry and employment around La Défense in western Paris. According to the Epamarne, this objective can legitimately be considered achieved since Noisy-le-Grand 'is currently the second largest centre of tertiary employment outside of Paris, after La Défense' (Epamarne/Epafrance, 1998: 2). To reinforce this, in publicity material emphasis is placed upon the preponderance of tertiary employment; the large amount of office space and favourable environment for busi-ness; the presence of large companies, the regional shopping centre and higher education and research establishments; the town's proximity to Paris and the good communications links with the capital via the A4 motorway and RER train link, which both pass through the town; and the easy access to both the Roissy-Charles de Gaulle airport to the north and the high speed train (TGV) station at Chessy to the east (Ville de Noisy-le-Grand, 2004).

One town or two? Employment and residential polarisation

Although valid to a degree, the image of a well-placed, modern commercial and business centre is not without its problems. In partic-ular, the new commercial and business centre was artificially grafted onto a pre-existing small town rural *commune* following the planning of the new town of Marne-la-Vallée. The result of this rapid emer-gence of a commercial and business centre in the decade after 1975 is a rather fragmented urban structure with two distinct centres, each housing different populations and serving different functions. On the one hand, there is the modern business district of Mont d'Est/Pavé Neuf with its large-scale modern and postmodern residential and commercial developments and, on the other, the Centre, which has retained the atmosphere of a traditional French small town or village.

The division between the old and new districts can be seen not only in terms of economic structures, but also in demographic structures.

At the beginning of the 1990s, 46 per cent of the population lived in the old town, and 54 per cent in the new. This represents a far greater density of population in the new town, as it makes up only 30 per cent of the surface area of Noisy-le-Grand. The new town is characterised by a younger population (38 per cent under 20, and 78 per cent under 40, compared to 26 per cent and 56 per cent respectively in the old town). Although the rate of economic activity is higher in the new town (52 per cent against 49 per cent) so is the rate of unemployment (10 per cent compared to 7 per cent and an average of 8.7 per cent for the whole town in 1990). Finally, the new town houses a higher proportion of immigrants (19 per cent against 11 per cent), although the gap is probably wider due to the number of second- and third-generation immigrants who have French nationality as they were born in France (Balaquer *et al.*, 1996: 10).

These differences are reflected in the housing stock: 93 per cent of accommodation in the new town is in the form of flats (*logements collectifs*), while in the old town 64 per cent is made up of individual houses; 62 per cent of households live in rented accommodation in the new town, whereas 71 per cent own their property in the old town; and social housing accounts for 46 per cent of properties in the new town against only 8 per cent in the old town (25 per cent in the town as a whole) (Balaquer *et al.*, 1996).

The Mont d'Est area has been successful in attracting major companies such as Groupama, IBM, RATP and so on. Indeed, at the end of 1999, Mont d'Est benefited from the presence of 47 per cent of all establishments employing more than 100 people in Noisy-le-Grand, and 386,600 m^2 of office space, out of the 450,000 m^2 in the town at the end of the 1990s (DDEE, 2000: 7). It is also the site of a 70,000 m^2 commercial centre, *les Arcades*.

Indeed, Mont d'Est, as home to the town's RER train link to Paris, the Arcades shopping centre and a large office complex, aims to serve not only Noisy-le-Grand but also the wider region, and has been designated a Regional Urban Centre (*Centre Urbain Régional*) for planning purposes. As the primary location for tertiary industries and services, and particularly of large companies, in the eastern Parisian region, it is thus the 'motor' district at the heart of future economic development policies for Noisy-le-Grand. It is also home to the residential complexes of the Palacio d'Abraxas and le Théâtre, a grandiose development of 590 flats completed in 1983. This accounts for half of the accommodation in Mont d'Est, and is largely made up of social housing – the 440 flats of the Palacio d'Abraxas being in the public sector (43 per cent of all housing in Mont d'Est is social housing).

The Centre is the second most populated part of Noisy-le-Grand after Champy, another new district which houses the higher education establishments that have been located in the town. The town hall as well as numerous small artisanal retail outlets and a covered market are located in the Centre, which also contains the prosperous residential area of the banks of the Marne, and the Espace Michel Simon, a $6000 \, m^2$ cultural centre containing a library and 715-seat theatre. Although Mont d'Est has a lower population than the Centre, when taken together with the adjoining Pavé Neuf it has a higher total than the Centre, now confirming it as the true town centre.

Self-containment

Noisy-le-Grand was originally conceived by the State as a multifunctional community within which people would live and work in the wider new town of Marne-la-Vallée and is in certain respects very different from areas of urban sprawl in Europe and certainly North America. Nevertheless, despite these intentions, Noisy-le-Grand shares some of the broader characteristics of post-suburbia. Perhaps more so than any of the post-suburban areas considered in this study, Noisy-le-Grand stands out as a space created by the State which continues to suffer from problems associated with a lack of place identity. According to one council official,

> We do not want to be a dormitory town for Paris ... We want to put in place all the services and employment structures necessary for a balanced life in Noisy: a hospital – which we do not have – educational establishments, work and leisure facilities so that people can find all they need in Noisy. The aim is to be a self-contained town which fulfils all the functions necessary in life for the inhabitants. [Interview N1]

Of itself, Noisy-le-Grand's lack of self-containment may not be unusual, but is symptomatic of a broader malaise in terms of identity.

An analysis of the daily migrations to work, of inhabitants, from 1999 census data reveals that only 20 percent of the 26,500 employed population resident in Noisy-le-Grand also work in the *commune* (compared to 30.5 per cent for the whole of the Ile-de-France region). Nearly one in three, 7300, are drawn to Paris for their employment. Even when the larger entity of the new town is taken into account only

13 per cent of employed residents of Noisy-le-Grand work in other *communes* of Marne-la-Vallée. On the other hand, only 5400 of the 24,000 jobs in Noisy-le-Grand (22 per cent) are occupied by inhabitants of the *commune*, with 18,500 occupied by outsiders, principally from the neighbouring *départements* of Seine-et-Marne (30.4 per cent) and Val-de-Marne (21 per cent), followed by Seine-Saint-Denis (17.3 per cent) (GEDA, 2003: 29–32).

Thus, Noisy-le-Grand is something more than a dormitory town or suburb yet not quite a US-style edge city in terms of its functioning predominantly as an employment centre. Employment is available, but is insufficient for the majority of the local active population. Such an apparently paradoxical situation can be explained by French land-use laws and the planning rationale behind the development of Noisy-le-Grand. By law, French *communes* must provide a certain percentage of social housing on their territory in the interests of social mix. This ensures a certain proportion of lower-paid workers in industry, construction and transport reside in the *commune*. In the case of Noisy-le-Grand, there are 5000 such workers, some 20 per cent of the employed population. On the other hand, as we have seen, the development of Noisy-le-Grand was predicated upon its role as a centre of tertiary employment within the eastern Parisian area. In particular, the attraction of headquarters of large companies to the locality over the past two decades or so has resulted in the creation of management positions within the town. These, along with journalists and university teachers, accounted for 27.9 per cent of employment in Noisy-le-Grand in 1999, but only 16 per cent of its active population (GEDA, 2003: 29–30). The result has been that there are only 3000 jobs for the 5000 industrial workers resident in Noisy-le-Grand, with a deficit also apparent for lower-level service-sector workers (8100 residents for 6000 jobs in the competitive sector, or 28 per cent of jobs and 33 per cent of the active population) and a surplus of managerial positions (6700 for 4000 residents in this category) (GEDA, 2003: 29–30). Thus, the general trend is towards the out-migration of industrial workers and lower-level white-collar employees towards Paris and other neighbouring centres of employment and for the in-migration of managerial staff (GEDA, 2003: 34).

Such a situation is not without its consequences. First there is a disconnection between the wealth created in the town and its distribution to the local population [Interview N11]. Second, there are implications in terms of transport in the locality, with traffic congestion being a major concern, especially for businesses in Noisy-le-Grand. In effect, while the majority (53 per cent) of Noisy-le-Grand residents use public transport to go to work, especially when travelling to Paris (82 per cent), half of those who travel to Noisy-le-Grand from elsewhere do so by car (GEDA, 2003: 29–30). The result is congested roads and problems with car parking. The town could provide employment for 80 per cent of its active population (GEDA, 2003: 30), thereby achieving the aim of providing a place where people can live and work without the need for travel between the two. As we have noted above, however, results have fallen far short of these ambitions with the vast majority of employed residents travelling elsewhere for work and a significant in-migration to fill those jobs on offer within the town. Thus, despite the existence of rational planning and strong State structures, Noisy-le-Grand displays some of the principal characteristics associated with urban sprawl in the US: a separation of living and working areas, the resultant need for daily migrations and consequent problems of traffic congestion, despite the locality being well served in terms of public transport.

6.3 A nowhere in search of an identity

More fundamentally, the administrative structures and geographic loca-tion surrounding Noisy-le-Grand further hamper the town's attempts to forge a distinct, positive identity. In effect, the town is bound into a series of interlocking administrative relationships which are the product of French arrangements for local government and of its special status as part of the new town of Marne-la-Vallée. Thus, Noisy-le-Grand is a *commune*, the lowest tier of local government in France, headed by a mayor and responsible for local planning matters and so on. It is also in the Seine-Saint-Denis *département* (equivalent to an English county), and part of the wider Ile-de-France region. While the former has no formal role in economic development, it does play an important role in infrastructure development. The Regional Council, on the other hand, has a major role in both infrastructure development and urban and economic development since President François Mitterrand's decentral-isation reforms of 1982. Some sense of these administrative complexities is captured in Figure 6.1.

Such administrative relationships are contractualised in a series of pluri-annual plans that guide the development of Noisy-le-Grand, as they do other *communes* in France (Cole and John, 2001: 51–4). Thus, the main strategies for the economic development of Noisy-le-Grand are contained in a series of overlapping projects and plans, principally the *Contrat de Plan Etat-Région* (CPER) signed by the Ile-de-France region and the regional Prefect, and the *Contrat de Ville*, signed by the towns of Noisy-le-Grand and Villiers-sur-Marne and by the central State in the form of the respective departmental Prefects.

The CPER is a 7-year planning agreement covering the Ile-de-France region, the basic aim of which is to ensure a balanced economic, social, urban and environmental development for the region. The most recent plan covers the period 2000–2006 and was heavily influenced by the State's regional planning body *Délégation d'aménagement du territoire et d'administration régionale* (DATAR), the *Institut d'Aménagement et d'Urbanisme de la Région Ile-de-France* (IAURIF) and by the Region's own administration (Ile-de-France Regional Council, 2000a: 5; 2000b: 5). Of the 59bnFF total expenditure, 20bnFF will be provided by the State, 30.6bnFF by the Region and 6.7bnFF by the local authorities concerned, primarily the *départements* (Ile-de-France, 2000b: 43). The influence of the central State in the formulation of the CPER can be seen not only in the role of DATAR but also in direct prime ministerial influence on the negotiating priorities and positions of the regional Prefect (Ile-de-France, 2000b: 21–34). Indeed, the main concerns of the central government appear as the primary points of focus in the final plan.

As far as Noisy-le-Grand is concerned, the first negotiating priority of the Prefect listed in the Region's report on the CPER is the economic renewal of the east and the north of the Paris agglomeration (Ile-de-France, 2000b: 6). Overall, the main priorities of the CPER are to improve transport, especially suburb–suburb, develop the research and university sector, employment creation and training and housing. Under the first heading, the Trans-Val-de-Marne tramway will be extended to Noisy-le-Grand and on to Champigny-les-Boulereaux in the east, thus improving connections with the new town (Ile-de-France, 2000a: 13). Marne-la-Vallée is also considered a priority as far as improvements in road transport are concerned, particularly with a view to 'improving the attractiveness of regional urban and economic development centres, and primarily of the new towns' (Ile-de-France, 2000a: 21). These works will be financed by the State and the Region on a parity basis (Ile-de-France, 2000a: 22).

As well as transport, the development of higher education and research in Marne-la-Vallée is also mentioned in the CPER as central to regional economic development strategies (Ile-de-France, 2000a: 41). Within the regional strategy for developing higher education, Marne-la-Vallée will receive a total of 106 mFF – 95 mFF from the Region and 11 mFF from the central State – to develop existing sites (Ile-de-France, 2000a: 53). The major operation concerning Noisy-le-Grand within the

CPER, however, concerns the construction of a multi-modal station in the town at a total cost of 140 mFF, of which the Region will contribute 124 mFF (Ile-de-France, 2000a: 14).

Part of the logic of the CPER for a balanced regional development is to concentrate action on certain areas that are seen as socially and economically underprivileged compared to the rest of the region with a view to reinforcing the global economic role of the region. In this, the CPER relies upon the laws of 25 June 1999 and 12 July 1999 rendering intercommunal co-operation stronger and simpler. Thus, in line with prime ministerial wishes, local action in job creation, social cohesion and sustainable development are to be articulated within, and seen as complementary to, regional and national development policies (Ile-de-France, 2000a: 93–5; Ile-de-France, 2000b: 22–6). This is envisaged as leading to a more coherent and rational intervention on the part of the public bodies concerned with putting the various elements of the CPER into action in a certain area, and leads, in priority areas, to a *Contrat de Ville* based around a partnership between local authorities (*communes* and *départements*) and the Region and the State. The CPER encourages, in particular, intercommunal *contrats de ville* and encourages partnerships with local public and private organisations. It also foresees the possibility of the State, Region and *communes* signing agreements for a *Grand Projet de Ville* (GPV) as a further complement to local and regional town policies. Inclusion as a site for a *Contrat de Ville* or GPV leads to additional finance (in the latter case for the most deprived urban areas of France) from both the State and the Region over and above those funds allocated to specific sectoral projects and policies (Ile-de-France, 2000a: 96–9).

Marne-la-Vallée is, in fact, one of the ten priority sites of the CPER given its 'regional and national interest [and] potential for development' (Ile-de-France, 2000a: 105). The State and the Region will therefore continue to play a major role in its development through a 'development plan' (*projet d'aménagement et de développement*) formulated in conjunction with the local authorities in place (Ile-de-France, 2000a: 105–7). In addition, within this new town structure, Noisy-le-Grand is the object of both a *Contrat de Ville* and a GPV.

In line with the orientations set out in prime ministerial circulars and taken up in the CPER, the *Contrat de Ville* associates two *communes*, Noisy-le-Grand and the neighbouring Villiers-sur-Marne, to form 'Les Portes de Paris'. The State is also a signatory to the contract – which covers the same period as the CPER, 2000–2006 – via the two departmental Prefects of the Seine-Saint-Denis and the Val-de-Marne.

According to the document, the agreement is the fruit of a consultation process involving State services (*Fonds d'action sociale, Caisse d'allocations familiales,* Epamarne), local authorities and employer organisations (*Chambre de Commerce et d'Industrie, Chambre de Métiers*) (Noisy-le-Grand and Villiers-sur-Marne, 2000: 4). The main aim of the agreement is to encourage and facilitate co-operation between the two *communes* and the State for the development of the area, and also to solicit the further collaboration of other partners including departmental level governments, either by getting them to sign up to the *contrat* or by signing specific agreements within its framework (Noisy-le-Grand and Villiers-sur-Marne, 1999: 5–6).

Spanning, as it does, two *communes* and two *départements*, the *Contrat de Ville* is innovative in France, and 'Les Portes de Paris' is very much a pilot site for this sort of project. Nevertheless, such a choice is not totally surprising; common planning arrangements have affected the two *communes* since the 1970s due to their role as the economic centre of Marne-la-Vallée, and the choice is in line with the strategic orientations of the 1994 *Schéma directeur*, which sees Marne-la-Vallée developing a European role, with, again, the 'Portes de Paris' as its economic centre.

The main priorities and orientations of the *Contrat de Ville* can be seen as closely articulated with those of the CPER. Indeed, the first *contrats de ville* were signed in 1993 and covered the period 1994–1998,[2] but were prolonged until 1999 so that they could be renegotiated by *communes* and the State within the framework of the 2000–2006 CPERs (*Noisy magazine,* 1999: 13). Thus, in the 'Portes de Paris', improving the transport infrastructure is a priority, especially suburb–suburb, the A4 motorway, the Trans-Val-de-Marne tramway and the construction of a multimodal transport hub in Noisy-le-Grand. Mention is also made of linking the SNCF station in Villiers-sur-Marne with the RER station in Noisy-le-Grand (Noisy-le-Grand and Villiers-sur-Marne, 1999: 7). All of these measures, however, escape local-level decision-making and depend upon the implementation of the CPER.

The picture in Noisy-le-Grand, however, is further complicated by its special status as part of the 'new town' of Marne-la-Vallée. Thus, the Epamarne was created as a local State body charged with planning a town spanning several *communes* and *départements*, with a view to producing a coherent whole of living and working areas linked by well-planned communications networks.[3] A decentralised arm of the State was seen as necessary to produce a rational and coherent urban planning policy, and the Epamarne therefore took on many of the planning and development functions normally within the remit of *communes*.

In the early phases of its development, the effect of this was to link the urban and economic development of Noisy-le-Grand to that of Marne-la-Vallée. Indeed, the former was to provide the motor for the latter within the framework of regional development policies which saw the development of the east of Paris in opposition to development of the west. However, such thinking on the part of Epamarne planners did not take into account the peripheral position of Noisy-le-Grand within the new town structure or the magnet effect of its strategic proximity to the capital city. Similarly, Noisy-le-Grand is on the southern tip of the Seine-Saint-Denis *département*, separated from the rest of this administrative entity by the Marne river and poor communications links. These administrative and geographic constraints pose a problem for Noisy-le-Grand as far as the identity of the town is concerned. Indeed, Noisy-le-Grand has been imposed as a State-created space that only partially identifies with its *département*, the new town and Paris [Interviews N10, N13]. Although close to central Paris, Noisy-le-Grand is at the boundary of Paris and an inner ring of *départements*. It falls within the Seine-Saint-Denis *département* (which extends northwards) and also borders onto the Val-de-Marne 'inner ring' *département* (which extends to the south and the west) and the Seine-et-Marne *département* (to the east) which forms part of the much larger Ile-de-France region (Figure 6.1).

First, then, there was a consensus among our interviewees, from both the departmental administration and the Epamarne, that Noisy-le-Grand 'is not attached to the rest of the *département*' [Interview N3]. As one interviewee commented,

> Noisy-le-Grand is a commune that is in the extreme south of the *département* that it is attached to – Seine-Saint-Denis – which goes right up to the north of Paris. So it's a town that is a little at the end of the world, that doesn't participate in *départemental* policies. That poses problems for the commune, which has difficulty establishing relations with the *Conseil Général* [the *département*-level council]. So it is isolated. [Interview N2]

Second, although the work of the Epamarne is all but complete in Noisy-le-Grand (it only intervenes in a few specific projects now and focuses most of its attention on areas such as the development around the Disneyland Paris site), Noisy-le-Grand's partial identification with the new town appears to be part of the problem of its lack of identity. On the one hand, 'I'm sure that nobody knows that Noisy-le-Grand is in Marne-la-Vallée. In addition it's in the Seine-Saint-Denis, so no-one could imagine that it is in Marne-la-Vallée' [Interview N4]. On the other hand, and for some purposes and to some audiences, Noisy-le-Grand is very much part of the new town. As one interviewee commented,

> Noisy-le-Grand is clearly identified with the new town. It plays on the proximity to Paris and the good transport links, but also on the image of Marne-la-Vallée. It doesn't play on the image of the [Seine-Saint-Denis] *département*. At a push, you could take it out of the *département*, say the Marne is the border. For a start there is the geographical aspect which makes the point: Noisy-le-Grand is a bit particular, it's not attached to the rest of the *département*. And Noisy-le-Grand is atypical of the rest of the *département*... [Interview N3]

Yet, there are significant ambiguities inherent in Noisy-le-Grand's being part of the new town. The Paris new towns have themselves suffered from a lack of identity. Moreover, standing at the western extreme of Marne-la-Vallée new town and close to Paris, Noisy-le-Grand's separate identity is confounded by some of its distinctive architecture that has been more closely associated with Marne-la-Vallée new town. For example, the *Arènes de Picasso* housing development otherwise known as 'les Camemberts' in Pavé Neuf and Le Théâtre (Figures 6.2 and 6.3) served

Figure 6.2   Central Noisy with DIAC offices and 'Les Camemberts'

Figure 6.3 Le Théâtre

for a long time as the symbol of Marne-la-Vallée new town (Dieudonné, 1992: 53). More recently still, Noisy-le-Grand's separate identity has been eclipsed by the vast Disneyland Paris development at
the eastern edge of the new town. As a result, any distinct identity
that Noisy-le-Grand has is lost in a conflation with its larger fellow

State-construct of the new town Marne-la-Vallée, as one interviewee observed.

> For anyone in Paris, Noisy-le-Grand is Marne-la-Vallée. And for French Parisians, Marne-la-Vallée is a small town with Disney in it. It's not Noisy-le-Grand... So there is a problem of the appropriation of the name, image and territory.... It doesn't stop people from living and developing here, but there is a problem of image. [Interview N4]

The Disneyland development at the eastern end of the new town is predicated on leisure and tourism from which Noisy-le-Grand can benefit only marginally. Most development in Marne-la-Vallée is now directed towards this eastern edge of the new town. Moreover, Noisy-le-Grand's role as the motor of the new town has, if anything, been eclipsed by Disneyland and its associated developments, thereby lessening the value of identification with, and implication in, the new town. In the light of the new easterly orientation, it is hardly surprising that tensions have become apparent in relations between Epamarnne and Noisy-le-Grand, with the latter recently expressing a desire to withdraw from the new town in order to complete neglected works scheduled in its territory. As the same interviewee neatly described,

> The town feels that it cannot get anything further from the Epamarne, so it is trying to get rid of an actor. Relationships between the different partners are very complex with a multiplicity of actors... each with its own terrain and a particular influence, with the result that you don't really know who is at the helm. [Interview N9]

Thus, Noisy-le-Grand's identity appears confused and its development is not seen as linked to its position in the new town of Marne-la-Vallée or to its place within the Seine-Saint-Denis *département*.

Third, the aspirations and promotional work of the municipality highlight a further partial identification of Noisy-le-Grand with the sprawl of eastern Paris. Whereas Epamarne literature emphasises the role of Noisy-le-Grand as the business centre of the new town, the municipality of Noisy-le-Grand markets the town as the 'tertiary centre of the eastern Parisian area' (*première pôle d'affaires de l'Est parisien*), with little reference to Marne-la-Vallée and none to the *département* (Ville de Noisy-le-Grand, 2003: 9). Here the municipality's identification with Paris appears to be based on several factors. There is, as we have seen, the relative isolation of Noisy-le-Grand from the rest of the *département* of

which it is a part. Such isolation is reinforced by relatively poor transport communications across the Parisian suburbs, as well as by a functional differentiation between Noisy-le-Grand and other *communes* on
the eastern edge of Paris. In effect, the Seine-Saint-Denis was – until its
tertiarisation over the last decade and particularly since the development
of the Plaine-Saint-Denis around the Stade de France, constructed for
the 1998 football World Cup – an industrial working-class area, whereas
the growth of Noisy-le-Grand was predicated upon the development of
tertiary activity. Likewise, although Noisy-le-Grand is supposedly the
economic heart of the new town of Marne-la-Vallée and is connected to
other *communes* within this structure by good road and rail communications, it is functionally separated from them. In effect, the economic
development of Noisy-le-Grand has been largely dependent upon large
employers relocating to the town, whereas the rest of the new town is
mainly dependent upon small- and medium-sized business.

The growth of Noisy-le-Grand, since the decision to develop the
commune was taken in 1971, has depended upon attracting investment,
employment and population from the Parisian core in order to create
a self-sufficient community spanning several *communes* in the new
town, with Noisy-le-Grand as its business centre. In order to achieve
this, public institutions and State-owned and private enterprises were
more or less forced to set up or relocate to Noisy-le-Grand through a
combination of planning restrictions in the Parisian core and cheap land
and office space. Rather than tying Noisy-le-Grand to the new town
as envisaged by the planners, however, such a strategy paradoxically
underlined the dependence of Noisy-le-Grand upon growth in the urban
core of Paris as the base for clients and institutional partners of inward
investors. Such synergy with the core has been reinforced with the development of transport infrastructures linking Noisy-le-Grand to central
Paris from the late 1970s onwards. The extension of the RER train link
from La Défense in the west to Noisy-le-Grand in the east means these
two tertiary centres are now only 30 minutes apart by public transport
and effectively part of the congestion and sprawl in the capital.

This symbiotic relationship with the capital city also appears to have
been strengthened due to economic changes that have occurred in the
north-eastern fringes of Paris. The Seine-Saint-Denis *département* has
de-industrialised attracting many companies in the audiovisual sector,
particularly in Montreuil and around the Plaine-Saint-Denis. With the
siting of the Ecole Louis Lumière (audiovisual school) in Noisy-le-
Grand, this has created possibilities for co-operation between Noisy-
le-Grand and other *communes* on the eastern side of Paris. Further

commonalities – based upon the belief that the west of Paris was privileged over the east in terms of investment via the State-Region contractual planning process and that the east therefore required more infrastructural investment, particularly in terms of suburb–suburb transport – led to the creation of the *Association des communes et territoires de l'est parisien* (ACTEP), a group of 17 *communes* and *départements* on the eastern edge of Paris. Noisy-le-Grand is the eastern-most *commune* in this informal association of local authorities created with the aim of promoting the economic development of eastern Paris [Interviews N8, N9, N13]. The leading role taken by the municipality within it – the Mayor of Noisy-le-Grand, Michel Pajon, was its first President from January to June 2001 – clearly demonstrates that the municipality sees the economic development of the town as inextricably linked to that of the eastern side of Paris.

In sum, then, Noisy-le-Grand's identity and development prospects are confounded by a wider set of non-local State projects and administrative relationships. As one interviewee neatly summarised, 'Noisy-le-Grand is all alone. It's trying to find a place for itself in what is going on around it' [Interview N2]. In Taylor's (1999) terms, non-local State practices have imposed Noisy-le-Grand as a post-suburban *space* – a space which has seen considerable physical development and is a sizeable centre of economic activity and population – and have also prevented local agents from investing it with a sense of *place*.

6.4   Embedding business: From space to place?

Noisy-le-Grand's lack of key services and its lack of self-containment are nothing especially unusual in the context of the other edge urban municipalities considered in this study. Yet it is a significant employment centre and so in this section we go on to consider the role business has played in urban politics and in shaping a sense of place identity.

Collective business bodies are few and have little influence [Interview N11]. For instance, after an initial period of activity the town's Club Ville-Enterprises (formed in the early 1990s) appears to have lost any influence it may have had. It now has more of a social function – its role being taken by the municipality's own Economic Development Department [Interviews N5, N8]. Major companies such as Groupama (Figure 6.4) belong to their relevant national and international professional associations but did not belong to the local Chamber of Commerce (with its focus on other parts of the *département*) or MEDEF (the French employers' organisation).

Figure 6.4 Groupama offices

Thus, although Noisy-le-Grand represents a quite sizeable concentration of business, these major businesses have very little attachment to, or dependency on (Cox and Mair, 1988, 1991), the locality that might engender participation in local political affairs [Interviews N12, N14] – a feature driven at least partly by the public, former public or quasi-public nature of several leading employers in the town. On the one hand, the State has obliged these companies to locate in a place where they would otherwise not be either by diktat or by refusal of planning permission to companies expanding or setting up within the capital [Interviews N5, N6, N7, N13]. So as an interviewee at one major employer noted, we 'set up in Noisy-le-Grand more than 20 years ago because the DATAR forbade us to expand in Paris, but proposed that we go to impossible areas such as the east of France, or here' [Interview N5]. On the other hand, some of the major employers were attracted as a result of the financial incentives that remained in force during the 1980s and 1990s. Larger companies appeared to evaluate relocation from Noisy-le-Grand on a regular basis with the 'social cost' of moving keeping them there for the time being [Interviews N5, N6, N7, N12]. As one interviewee elaborated,

> We aren't embedded in the town. We are here, we live here, our staff come to work here, but it could be Noisy or anywhere. Honestly, we regularly ask ourselves, and have done recently, whether we wouldn't be better off elsewhere. [Interview N7]

Companies perceive that they are embedded in Noisy-le-Grand, not through any formal influence they may have in local political decision-making, but by virtue of the fact that their employees live locally and in *communes* within commutable distance [Interviews N5, N6, N7]. In practice, as one town hall official remarked, this means that as far as the economic and physical development of Noisy-le-Grand is concerned, 'We are in the hands of estate agents and the State' via the Epamarne, in partnership with the local authorities and the developers to whom office blocks are sold or leased once they have been constructed [Interview N1].

One of the consequence of this is that until recently the physical development of Noisy-le-Grand has been dominated by large-scale commercial developments, with a proliferation of modern office blocks tailored to meet the needs and interests of the large companies that the town attempts to attract in order to heighten its prestige as a tertiary centre, particularly around the Arcades commercial centre in the Mont d'Est area of the town. Despite a deliberate mixing of economic activity and residences to integrate Mont d'Est with the rest of the town, it has remained mainly a business centre rather than a 'true' town centre due to a lack of social animation in the evenings and weekends (Balaquer *et al.*, 1996: 60). The town has attempted to overcome such problems through the promotion of local festivals and projects for the embellishment of the town – financed in part through GPV monies, including a pedestrianised link from the Centre to the Mont d'Est area and landscaping of the Marne river banks – in order to produce a more cohesive urban whole (*Noisy magazine*, 2001b). In addition, leisure activities are also being developed, notably through the construction of *Libercité*, a leisure centre with France's first indoor real snow ski slope (DDEE, 2003b).[4] From being 'pioneers', the population of Noisy-le-Grand appears to have grown attached to their locality, residing and consuming in the town, even if they do not necessarily work in it.

The limited local dependence of industry has, from the outset, been conditioned by the State itself. It results in an 'individualistic' company outlook as one interviewee identified

> companies in Noisy and Paris do not belong to networks and things like that. They are concerned with their economic activity.... there are no links between companies. We don't have the time to organise anything, to meet. It takes time... You mustn't forget that the inter-locutors of the State are the local authorities, the mayors, the *députés* (MPs), not companies.... So in general, businesses have little influ-ence. [interview N5]

Instead, the preferred channel of influence for large companies as far as local economic development is concerned is through direct access to the Mayor and his senior officials [Interviews N5, N6, N7].

Conceived in narrow terms, the privileged direct access accorded by the Mayor to major businesses in Noisy-le-Grand is partly a function of prestige and the prosaic realities of the fiscal [Interviews N5, N6] and broader economic contribution made by these larger companies and hence their impact on the budgetary stability of the municipality. Such influence, however, must be weighed against the political priorities of the Mayor and his constituents as his need for votes means that he cannot listen only to business, even of the large companies on his territory. Business views must also be weighed against State funding priorities in a place such as Noisy-le-Grand, where State expenditure plays a major role in local economic development. The result is a certain level of informal influence over local matters – which at times needs to be exerted through threats to leave – but little formal influence.

Conceived in broader terms, the pattern of business interest represent-ation has not transcended the sorts of administrative structures that have imposed Noisy-le-Grand as a State-created space. The geographical posi-tion of Noisy-le-Grand means that it is isolated in Seine-Saint-Denis. As the activity of the Chamber of Commerce, for example, covers the whole *département*, and tends to be focused on the regeneration of the north of the area, around the Plaine-Saint-Denis, it has little to offer local business in Noisy-le-Grand other than indirectly. There are also beliefs among the private sector in the separation of the private and public spheres [Interviews N10, N11], and an acceptance that the democratic process means that interests other than those of business need to be taken into account when planning decisions are made [Interviews N5, N7].

Companies in Noisy-le-Grand have voiced concerns and have had an impact, but only on highly localised issues – those related to the insecurity of car parks and the lack of parking places [Interviews N8, N10]. Even though transport is considered to be a major problem due to traffic jams around the town, and particularly on the A4 link to Paris, businesses' lack of involvement mirrors the municipality's own struggles to assert itself within administrative circles. Transport is seen as a regional problem, affecting the east of Paris, and the whole Ile-de-France region, and as such is one that needs to be addressed by the public powers, which have a responsibility for creating a good environment for business. Pressure from business for the resolution of such problems is transmitted via mayors so that improvements can be made via public action, through the Regional, Departmental and Town Plans signed between the local authorities and the central State [Interview N13]. As another interviewee confirmed,

> the Mayor negotiates all that with the State or the appropriate ministry. The Mayor through informal exchanges, when we go to see him, hears the needs and demands of business and puts his own political sauce on it and negotiates with the State. He takes the demands of business into account as far as is possible. [Interview N6]

Being also an MP, the Mayor has access to national decision-makers and State funding. The complexity of local development structures – with the many layers of local, regional and national government involved in the capital region, as well as the Epamarne planning agency in Marne-la-Vallée – makes the Mayor a pivotal figure in an institutional maze and a natural point of contact for local business. However, the effectiveness of mayoral political manoeuvring appears to have been constrained along party-political lines as one interviewee, speaking in 2000, explained.

> Before, the Mayor was from the Right and the State was run by the Left, and this led to failure at the local level. Now we have a socialist Mayor, Michel Pajon, the region is in the hands of the Left and so is the State. This means that we obtain more for the town. [Interview N1]

Moreover, even when effective, Noisy-le-Grand's origins in State planning and its entanglement in a web of wider administrative relations
has locked the Mayor into a discourse centring on the neglect of the
town by, and appeal to resources from, the State [Interview N9].

The result, depicted in Figure 6.5, is a very fragmented business lobby,
which seeks influence through direct channels of communication to the
local Mayor, and one which is mainly confined to local issues rather
than wider issues of strategic planning. Business views are received via
public-sector interlocutors and are several times removed from decision-
making arenas that are embedded within what is a complex hierarchy

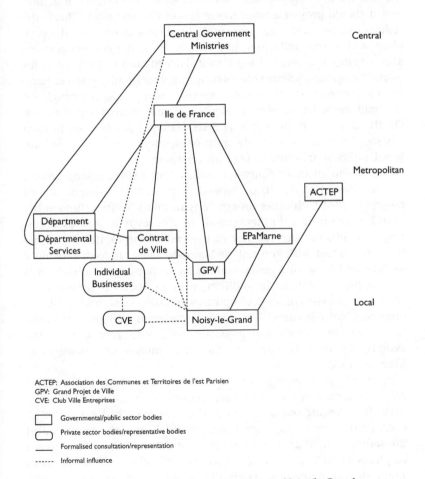

ACTEP: Association des Communes et Territoires de l'est Parisien
GPV: Grand Projet de Ville
CVE: Club Ville Entreprises

☐ Governmental/public sector bodies
⬭ Private sector bodies/representative bodies
── Formalised consultation/representation
------ Informal influence

Figure 6.5 Map of business interest representation in Noisy-le-Grand

of administrative arrangements, and as a result their direct influence within the edge urban political arena is muted.

6.5   Conclusion

From a rural *commune* on the eastern fringes of the capital, Noisy-le-Grand developed into a medium-sized tertiary centre over a short space of time. Such rapid growth and development was very much a State-managed affair. However, even where States rationally plan human settlement, they cannot control all the parameters of development. On the one hand, Noisy-le-Grand has not fulfilled the original intention that it should provide a town where people live and work. This is the result of many factors: its geographic location – at once on the periphery and in proximity to, if not directly on, major communications axes – makes it a place of significant daily in- and out-migration for work. Contradictory State policies concerning social mixity and tertiary-centred economic development exacerbate such problems through the resultant mismatch between residents and employment opportunities. On the other hand, neither have State planning practices set in train any significant autonomous dynamic of identity formation in the local social, political or business life of the *commune*.

The development of Noisy-le-Grand, rather than checking growth in Paris, would appear to be dependent upon it. The municipality's rapprochement with other eastern Parisian municipalities through the ACTEP suggests that further growth in the town is seen as occurring in symbiosis with the rest of the east of the capital city rather than in tandem with the rest of Marne-la-Vallée. Indeed, proximity to Paris and increasing commonalities in economic profile make this a coherent orientation. In addition, with its concentration of large companies, modern office developments and large social housing developments, Noisy-le-Grand has more in common, not only economically, but also socially and physically with other *communes* on the eastern edge of Paris than with the other, more rural, *communes* of Marne-la-Vallée.

The result, we have suggested, is the creation of a town that has had difficulty in creating a sense of place identity. Noisy-le-Grand is 'finding itself. It is coming out of a period when the State had a great influence, particularly through the Epamarne. In a sense they have "killed the father", but that is not enough' [Interview N10]. A new recent emphasis on urban renovation and the embellishment of the town alongside continuing projects for office construction may go some way to correcting this but what Noisy-le-Grand lacks for a town of its size are social amenities. New projects such as that for *Libercité* indicate that Noisy-le-Grand may well be entering a new phase of development. The town is still a pioneering one, but the accent now is on fashioning a true living place out of an administratively created working space.

7
Espoo: California Dreaming?

This is like California. You need two cars. If you don't have two cars you are in trouble.

Interview E8

7.1 Introduction

Many of the most salient features and some of the emerging contradictions of, and inequalities associated with, the very rapid and very recent urbanisation that has taken place in Finland are distilled in the urban politics surrounding the growth of the municipality of Espoo which stands to the immediate west of Helsinki (Figure 7.1). Castells and Himanen (2002) have recounted the story of one apparently paradoxical progeny of Finland's strong welfare state system – namely its coexistence with a highly internationally competitive information technology industry (Van den Berg *et al.*, 2001). The story of the growth of Espoo, itself now one of the major concentrations of the information technology industry in Finland, reveals another paradox – namely the coexistence of localised American-style processes of urban development with a strong national welfare state framework.

A large, and once entirely rural, municipality, Espoo has grown from the early post-Second World War years into Finland's second largest city and would be rival to the nearby capital city of Helsinki. As recently as 1950, Espoo had a population of just 22,800. This had grown to 221,600 in 2003 (City of Espoo, 2003). Although the municipality has become known for its information technology industries and is associated with Nokia in particular, the dynamism of the information technology sector and the bulk of its employment within the municipality is provided by new small businesses rather than the bigger companies. Espoo is home to

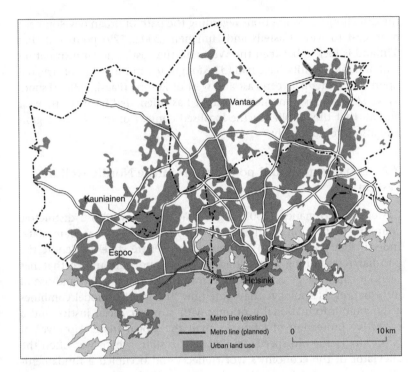

Figure 7.1 Map of Espoo

the headquarters of a number of major Finnish private-sector companies (including Nokia and Fortum) drawn from a range of industries although it is public-sector enterprises (including Espoo City and Helsinki University of Technology) that are also among the largest employers locally.

As with several of the post-suburban municipalities in this study, Espoo can lay claim to a long history of settlement. The City's own literature proudly claims a lineage that dates back 9000 years to the first inhabitants of the area. Of more significance is Espoo's ecclesiastical and regal role as a staging post along the King's Road from Turku to Viipuri dating from the thirteenth century. As a result, the oldest building in Espoo is its church which dates from the 1490s (City of Espoo, 2003).

These heroic appeals to historical identity may seem arcane; however, they have a tropic resonance, perhaps more so than in any of the other municipalities considered in this book, with one of our major themes – namely the value of comparative study to a fuller understanding of variations on common themes of post-suburban form and function and the dispelling of assumptions regarding the path of urban development from east to west. Castells and Himanen (2002: 129) point out that 'Finland is locked between the West and the East'. The Finnish nation and national identity have had to be fought for and socially constructed upon the border between east and west of Europe (Paasi, 1996). Espoo, as we shall see, is a municipality that has orientated itself to the west and all that this implies in the localised context of the Nordic region and also internationally.

7.2   Urbanisation and polarisation in the Nordic welfare system

Castells and Himanen (2002) argue that there has been a distinctive coupling of the information society with the Nordic welfare state. Indeed, they use the term 'informational welfare state' to denote the peculiar refashioning of the traditional welfare state model that has seen Finland emerge as a leading centre for the production and use of information technology. They note how 'the Finnish model combines a dynamic informational economy with stronger social justice and a collective protection of labour... than the global trend. However, it is no longer the old species of the welfare state, which was often the alleviator of the economy's worst effects and occupied a fundamentally defensive position against the economy' (Castells and Himanen, 2002: 87). The Nordic countries have 'a strong social democratic value consensus, which, unlike Britain for example, has been generally little disturbed over the last twenty years' (Goldsmith and Larsen, 2004: 122). The Finnish welfare state has retained this consensus and its legitimacy by casting the pursuit of the information society as one in a sequence of challenges to Finnish survival. Moreover, it has bolstered the growth of information technology industry by promoting social uses of information technology (including a renewal of welfare state institutions), and, in several different ways, creating an intermediary role for new state institutions in the innovation process.

To date, this coupling of the Finnish welfare state to information technology industries has, as Castells and Himanen (2002) note, produced very low levels of social polarisation compared to most developed nation economies. However, Vaattovaara and Kortteinen note there is a clear turn towards urban differentiation associated

with the new nature of economic growth after the depression: it seems that growth based on ICT technology and a globalised

economy, or informational economy... breeds urban inequalities
even in political conditions specifically designed to prevent this from
happening. (2003: 2130)

As such not only is the information society centred on the Helsinki
metropolitan area within Finland but there are emerging signs of social
differentiation emerging within this region. Pre-figuring Florida's (2004)
recent elaboration of the geography of the creative class, Illmonen *et al.*
(2000) have shown how the residential preferences of different segments
of the creative class have diverged markedly within the Helsinki metropolitan region. Design professionals have a preference for the inner city
environment of Helsinki whereas those employed in Finland's dynamic
information technology industries choose a suburban environment
outside Helsinki. This is a picture that is also painted by Vaattovaara
and Kortteinen who describe how

> The information sector has become the main engine of the whole
> region. In addition, it seems that this development is clearly linked
> to the new spatial differentiation: most firms of the new information
> sector have settled in the western parts of the region, surrounding the
> Helsinki University of Technology, either in Espoo or in the western
> parts of the centre of Helsinki, especially Ruoholahti. (2003: 2135)

These preferences for different residential locations are associated also
with strikingly different urban forms and housing tenure. The greater
density of Helsinki city and the dominance of rented apartment blocks
can be contrasted to the greater prevalence of owner-occupied detached
or row-houses in Espoo.

The first major development in Espoo was the Finnish garden suburb
or city of Tapiola. Initially at least it was intended that Tapiola would
represent a new way of living for a cross-section of Finnish society rather
like the British garden cities and new towns. However, Tapiola quickly
became associated with a degree of affluence unusual in Finland. So,
for example, 'many who received a new flat in the quality-designed
houses in Tapiola saw it as equivalent to winning in the lottery' (Tuomi,
2003: 26). That the affluence of Tapiola had become apparent by the
1960s was underlined in the title of Ossi Hiisio's polemical book entitled
Tapiola – the Village of Better People. This was due in no small measure
to the Arava housing system which, although open to all, effectively
selected in favour of wealthier Finns, since at this time it also required
the availability of some savings [Interview E1]. Prior to 1960, the Arava

system prohibited the re-sale of houses in order to prevent speculation. However, after 1960 the original purchasers were able to pay off the state loans and sell the houses which commanded high prices within the Helsinki metropolitan setting (Manninen, 2003b). The different urban forms that exist in Helsinki and Espoo have been *produced* from contrasting micro-social models.

The renowned Finnish architect Eliel Saarinen presented the first plan for the Helsinki region which as early as 1915 identified parts of the municipality of Espoo, such as Leppävaara and Tapiola, as locations along major axes of future expansion for Helsinki city. This plan spoke to the integrity of a greater Helsinki prior to the major urbanisation and industrialisation in Finland. However, whilst this may have represented the first coherent plan for the future growth of the Helsinki city-region it also apparently 'presented an idealistic vision of the city as well as being a speculative building project aimed at making a high financial profit' (Bell and Hietala, 2002: 169). In this respect it prefigures the *process* of urban development that was to occur in municipalities neighbouring Helsinki City – a process that has often been at variance with any unifying planning and administrative discourse regarding the city-region.

Herrschell and Newman (2003) suggest that 'Finnish regional governance... is based on a competitive dualism between strong local autonomy as part of municipal self-government and direct central government engagement through strict guidance, particularly in local government's role in welfare provision, education and healthcare' (2003: 77). Despite the creation of YTV – a metropolitan tier of government for the Helsinki city-region – this autonomy of municipal government has promoted differing stances or micro-social models within the capital city-region. Thus, 'Espoo and Vantaa have not been very eager in this co-operation but they saw that if they do not cooperate on a voluntary basis then the state will make a law and put them together' [Interview E3]. Swedish Crown land was handed over to the Finnish state or else to Helsinki city which is unique in Finland in owning most of its municipal territory.[1] The importance of this within the Nordic welfare state model could hardly be overstated. As one interviewee described,

> Espoo doesn't have anywhere near the same amount of land [in its ownership]. So within one metropolitan area we have a severe contradiction that two models operate. One being the Helsinki compact garden city approach where eighty-five per cent of people live in apartment blocks, only four per cent have access to terraced

housing and just eight per cent live in detached or semi-detached houses... We also have a key element in the plan to extend the metro... [to] Espoo which is very much an American model based on the private car. [Interview E2]

Levels of car ownership and usage in European terms are high in Finland since this is the major means of long distance travel between urban centres. Within the Helsinki metropolitan area, however, there are also important differences in the level of car ownership and usage among municipalities with Espoo residents making considerably greater use of the car as a mode of travel than residents in Helsinki (YTV, 2002). These elements of divergent social models existing within the Helsinki city-region prompted the Organisation for Economic Co-operation and Development (OECD) to observe that

The Helsinki Metropolitan Area has done an exemplary job in planning livable and affordable communities that are well-served by infrastructure and that contain well-designed workplaces. Projects like the technology park in Espoo, however, could be harbingers of problems ahead. Although its buildings are designed and constructed well and they are located adjacent to the Helsinki University of Technology, the physical paradigm is one of more suburban location and layout that leads to growing automobile usage and dependence.... It is much like an American-style 'office park', with large surface parking lots, broad building set-backs, curving access roads and a relatively narrow mix of land uses. (OECD, 2003: 84–85)

Aspects of Espoo's urban form do indeed resemble something approximating an American suburban office and retail landscape. There are 'some problematic practices lingering in Finland from Modernist architecture and town planning.... There is increasing dependence on and use of the automobile... Hypermarkets and shopping malls with aggressive signage and large parking lots are becoming the norm in parts of Vantaa and Espoo' (OECD, 2003: 93). In name, the newly built 'Big Apple' shopping centre at Matinkyla makes direct appeal to American culture while office developments also add to the edge-city-like feel in places of this large municipality.

There is a further irony here in that the same garden city ideal or, more correctly, aspects of it have produced these divergent compact city and garden suburb outcomes. A fundamental ingredient of Ebenezer Howard's garden city ideal was the ownership of land. As already noted,

Helsinki is unique in the Finnish context in owning the majority of land within its territory, which in itself greatly improves the efficacy of the city government's structure planning. Moreover it has aggressively acquired land in neighbouring municipalities Vantaa, Espoo and Sippo to provide for the expansion of its population long envisaged in a sequence of structure plans. In so doing, and with greater administrative resources at its disposal, it has also in the process tried to impose elements of its own compact city form on these municipalities. Yet, the integrity of Tapiola garden city was also ensured by the developer owning the entirety of land from the outset. Tapiola was the template for a patchwork of developments that were in themselves planned but yet created an unplanned sprawl in the long-term absence of any effective overarching structure planning or politico-administrative will to achieve it. It is to a consideration of these utopian roots of Espoo's urban sprawl that we now turn.

7.3   Gardens of dystopia? The planning of Espoo

The roots of the modern Espoo that has become Finland's second largest city lie in the *process* of development rather than the aspirations crystallised in the classic Finnish garden city – Tapiola. In this respect, the dispersed pattern of urban development in Espoo has clear parallels with the sort of planned sprawl experienced in the US (Gottdiener, 1977; Hise, 1997). This process of development in turn was something experienced more generally in the urbanisation of Finland but most acutely in the capital city-region. As Sundman describes,

Local large scale builders had appeared just before the war in the Helsinki region, in the shape of housing companies geared toward the production of social housing. In their wake came the private commercial builders . . . Massive investment in housing helped accumulate capital in the building sector, which in turn made it possible to purchase large tracts of land particularly in the region of the capital city . . . Planning capacity and the technical knowledge needed for realizing large-scale developments were available in the old towns, but not in the most expansive areas round towns. Under these circumstances the initiative for planning and building houses, as well as schools, nursery schools, roads, water mains and sewers slipped into the hands of the private building companies and land-owning interests'. (Sundman, 1991: 92)

Certain specific antecedents of this process of development can also be traced back to at least the late 1800s when Lars Sonck published *Modern Vandalism: The Town Plan of Helsinki* in which he criticised the desolate appearance of central Helsinki. The result was an architectural competition, the like of which has become a characteristic of urban development in Finland (Sundman, 1991: 71 and 112). Tapiola itself was not developed until the 1950s as a garden city and yet, in the specific post-war and municipal context, set a precedent for a patchwork of developments across Espoo that constitutes the decentred, urban sprawl that Espoo has become.

Brave new world: The building of Tapiola Garden City

The first land-use planning legislation of significance in Finland dates from 1932 and was an inheritance from Swedish rule. This act was a response to uncontrolled development in and around major cities and was an attempt to regulate economic relations between private land owners and communities (Sundman, 1991: 80). The next major piece of legislation again sought to limit the possibilities for uncontrolled development by granting municipalities the monopoly on planning but only dates from 1957 and did not come into force until 1959 – well after some of the key developments had already begun to shape the process of development in Espoo. From 1905 to 1960 no municipality had been granted town status. During the 1960s and 1970s the regulatory position of many former rural municipalities, including Espoo, was enhanced as the number of these classified as towns increased (Sundman, 1991: 92). Moreover, Espoo City itself did not form a planning department until 1974. The first master plan for Espoo was approved in 1978, effectively endorsing a pattern of development and rationalising it in terms of a vision for a multi-centred city [Interview E4].

Post-war Finland faced a chronic shortage of housing and this was nowhere more apparent than in the Helsinki area. Not only was Finland a predominantly agrarian society starting on the path to rapid urbanisation and industrialisation, much of which would be focused on Helsinki, but there was a need to house returning service men and refugees from the Second World War. Finally, as these developments gathered pace, planning also had another distinctive and ironic contribution to make to the form and process of urban development in Espoo. During the 1960s and 1970s, forecasts overestimated the population growth in municipalities such as Espoo which was being predicted to grow to 340,000 inhabitants by 2000 – thus sanctioning the hasty and fragmented developments in this land-rich municipality [Interview E4].

In the 1950s, Espoo was a predominantly rural Swedish-speaking municipality composed of a number of larger farms or manors. The only notable new settlement of the municipality had begun gradually as the wealthiest of Helsinki residents established summer houses along its lakesides and shoreline. The Finnish state had organised the compulsory parcelling of private land held in larger farms and estates in rural municipalities such as Espoo in order for ex-service men and refugees from Carelia and Porkkala – areas lost to the Soviet Union. These parcels of land carried the right to build. They were accompanied by a parallel private parcelling of land in some of the larger farms and manors in response to a continuing demand for single family dwellings. In the context of weak town planning powers, rural municipalities such as Espoo had little control over this fragmentation of landownership and associated building.

In 1952, in this context of a severe housing shortage and fragmentation of local landownership, the Finnish Population and Family Welfare Federation arranged a competition for the development of housing design and production. The economics of house building were important considerations in this competition, including an emphasis on the standardisation and prefabrication of housing production. The Family welfare Federation had appointed Heikki von Hertzen to oversee the competition and to realise elements of it. By the early 1950s, part of the large Swedish-owned manor of Hagelund at the extreme south-east of Espoo had already been sold to the Finnish state for the building of Helsinki University of Technology at Otaniemi. Moreover, the owner had also commissioned Professor Otto Meurman – the first University Professor of Town Planning in Finland – to make a plan for the development of the remaining land. When it became apparent that this remaining land with the existing plan was for sale, Von Hertzen moved to secure finance and create Asuntosäätiö – now a major regional housing association (Manninen, 2003a).

Otto Meurman's plan for the Hagelund estate became the initial plan for the new settlement of Tapiola. The original plan was influenced strongly by Ebenezer Howard's ideas of garden cities. Yet Tapiola represented a distinctively Finnish version of the template provided by Ebenezer Howard. As one interviewee described,

Maybe we are used to living very near to nature. When we are talking about garden cities in the Howardian way, so we are talking really about gardens but in the Finnish versions we have original nature combined into parks, gardens and small parcels. And this combination is typical for Finland. But there are many different versions of the garden city in different countries. In Tapiola, this idea of Howard's . . . about the ownership of the land has been realised in some ways. [Interview E1]

Some element of the Finns' greater incorporation of nature into garden city planning can be seen in Figure 7.2. Here, Von Hertzen added his own inflection and marketing to Meurman's original plan when posing the rhetorical question – *Koti Vaiko Kasarmi Lapsillemme?* (Homes or Barracks for Our Children?). In this way the aspirations for Tapiola garden city were set squarely against the urban form of central Helsinki (Figure 7.3). These sentiments were proudly proclaimed on an information board found in the centre of Tapiola during the early years of construction: 'Tapiola is not only a suburb. It is a miniature community. . . . Man is an essential part of creation and his connection with nature should not be severed. Barrack-like housing and miserable backyards are poor solutions and therefore Tapiola became a garden city' (Tuomi, 2003: 6).

If the aspirations for Tapiola garden city were borne of an earlier era associated with Ebenezer Howard, the ideas generated from the competition formed a backdrop of a shift among architects and construction companies towards a high point of Finnish modernism. Against this,

Figure 7.2 Low-density housing in Tapiola

Figure 7.3   Apartment blocks in central Helsinki

the original garden city vision of Meurman for Tapiola was modified
(Tuomi, 1992). In fact, Meurman disengaged himself from the planning of Tapiola while von Hertzen applied his energies in the vigorous
marketing of the new city. As one interviewee describes,

> This area was marketed quite heavily in an American style by Heiki
> von Hertzen and Asuntosäätiö. And people realised that there is some-
> thing special coming. And people who had different backgrounds
> understood that this is a new area that could be nice to go to ... They
> were really brave people that moved to a new area about which they
> only knew the plans. [Interview E1]

Moreover, something of the nearby development of Ottaniemi into the
campus for Helsinki University of Technology must have rubbed off on
the planning and building of Tapiola. Alvar Aalto – perhaps Finland's
greatest exponent of modernist architecture – was centrally involved
in the Ottaniemi development. While in the US, Aalto had worked
for the idea of 'the American town in Finland' as a model for urbanisation and was important 'in designing the strategy for the post-war
renewal programme and in maintaining a strong social and humanistic
commitment to developing the architectural basis for industrial mass

production' (Sundman, 1991: 86–87). These aspirations found expression upon the blank rural canvass that was Espoo. If 'Helsinki has the appearance of a planned city with only minor problems created by uncontrolled urbanization' (Laakso and Keinanen, 1995: 121), Espoo has come to represent a rather different pattern of urbanisation as one interviewee described.

> Then in Espoo because the culture is different and maybe the history too... In Espoo the city is not a very big land owner ... That's why the history of Espoo has been that private land owners have made agreements with Espoo about bigger areas and plan and build areas... In Espoo, town planning has been at another level... because they have not so many workers there and they don't have money for the infrastructure and so on, they have always handed things over to developers.... But maybe... because Espoo is a new city the people have moved from elsewhere in Finland and it has been very necessary to make planning such that it is possible to build. Without these private developers and builders Espoo would be a very small city. [Interview E5]

Utopian roots of an American-style dystopia

There is an irony in the 'planned' origins of urban development in the formerly rural municipality of Espoo. Despite the first major urban development in Espoo having been thoroughly planned as a garden city experiment, the subsequent pattern of development was essentially unplanned. As one interviewee described, 'from the outset, after the second world war, Espoo never really had a clear vision of what kind of a place they would like it to be. It was a very rural area. Even in 1954 when Tapiola garden district or city was being built it was considered quite far from Helsinki. Today it's considered almost to behave like a dormitory suburb' [Interview E2].

Heiki von Hertzen's descriptions of the development of Tapiola reveal the rural, unsophisticated nature of the municipality in which Tapiola was to be built. At the outset then, the city fathers were uninterested in Tapiola, as he recounted.

> We in Espoo have for almost five hundred years followed the principle that each house has its own well and... privy. The system has worked excellently for hundreds of years and will work this way in the future too.... You can of course, build a garden city here, we won't oppose

it as long as you pay for it all yourself. You must build the streets, the sewer systems, the sanitation department, the water pipes, the street lighting. . . . You are allowed to do it, but you won't get a penny from the municipality. (quoted in Nikula, 2003: 116)

As such, the municipality which had not been party to the sale of land also abdicated from any responsibility for the planning, developing and servicing of Tapiola under an agreement signed with the developer in a pattern that became familiar in the following decades in Espoo. In effect, the municipality was the outsider in a rapid, haphazard process of development as one interviewee described.

when Asuntosäätiö planned and built Tapiola, Espoo was the outsider. It was very simple for the developer . . . without some kind of regulations . . . and the people came from northern Finland to work here. They didn't look after the quality of the urban areas and so on. They built the row houses and high rise houses as much as possible without thinking about the quality and living environment and so on. . . . The builders . . . did down the image of these housing areas that now there is a very big work to change the image of those areas. [Interview E5]

The municipality was the outsider in these agreements with developers and at this time was in a weak position to affect the pace and form of development. Indeed in the absence of a planning department in the city Council, land owners or developers themselves effectively made detailed plans for the developments for which they were seeking council approval. So as another interviewee identified,

When you think of the other places, Matinkyla and Espoonlahti . . . So they are really some building companies which have bought the land and then Espoo City has made agreements with them. The developers have actually chosen how the development would happen. . . . in a way I don't think the decision makers in Espoo had enough power to change the development. The real estate policy here has not been as strong as in Helsinki. [Interview E6]

Moreover, although the initial developments in Tapiola were microplanned as coherent wholes usually overseen by a single architect, the new garden city began to expand through architectural collaborations which only conformed loosely to Meurman's outline plan and which were increasingly driven by commercial expediencies. Nevertheless a distinctive 'urban structure . . . gradually emerged from the "unplanned" planning of these collaborative architectural efforts' (Sundman, 1991: 90) – a form of urban development that has characterised Espoo more generally.

From the 1970s the city began to present itself as a city of five district centres. This principle was formally enshrined in planning documents from the late 1970s [Interview 4], although it was borne out of a necessity for planning to reflect the actual pattern of urban development [Interview E6]. This manner of development represents an ongoing burden of service provision with attendant fiscal issues, as the same interviewee described.

> If this had been developed or planned as a master plan you would never have opened certain areas there without any idea of who is going to pay for the infrastructure. I think it is from the 1950s and 1960s, they opened up all these housing areas where the single family houses are spread all over but what have been the consequences for infrastructure? It was enough to get this plan and they got their own water and waste management. So they didn't care about it. And today we are facing the problem. . . . It was calculated that we are lacking about 170m euros which we need to build infrastructure for those older areas with the single family houses. [Interview E6]

Plans in Espoo have therefore followed development rather than constraining or directing it – a pattern not unfamiliar in municipalities adjacent to established towns and cities. Indeed 'the community structure was so fragmented by the building of the 1960s, that subsequent overall land-use planning was seen as a way of "filling in" and "making whole"' (Sundman, 1991: 93). Espoo provides a striking case in point since, as one interviewee noted, 'the planning process and people trying to make services for people are working all the time like a fire brigade because the growth has been so very big over time' [Interview E1].

In the 1960s an isolated development in Vantaa was incorporated into Helsinki City – a phenomenon that, at the time, seemed a natural outcome of the outward expansion and superior resources of Helsinki City. Large tracts of land in Espoo (e.g. Leppävaara) have been owned by Helsinki City and developments that occurred there after Tapiola were the subject of possible incorporation into Helsinki City boundaries [Interview E3]. As the balance of forces among municipalities has shifted, Helsinki City has, until recently, prevented the further development of housing and amenities in the area. Only in the last decade as

Helsinki City's tax base has deteriorated has it relented in an agreement which allows Espoo City to develop the Leppävaara area with the former supplying the services and paying for the privilege [Interview E7].

Indeed, the lack of involvement from Espoo municipality in service provision for Tapiola meant that, initially, it was considered a residential suburb of Helsinki that might be incorporated into Helsinki City boundaries. 'In the beginning of the 1950s there was a general belief that Tapiola would become part of Helsinki, and the plans of the first stage included, for instance, a tram line running from the capital to Tapiola' (Tuomi, 2003: 13). And as one interviewee elaborated, 'for private organisations also here in Tapiola it was a better prospect in the future to be part of Helsinki than to be part of what in those days was a rural municipality' [Interview E3]. Since its development, there have been proposals for the incorporation of Tapiola into Helsinki, and indeed for a metro or tram line link, into Helsinki City.

However, by the late 1960s the outlook of the municipality in Espoo had changed towards one which saw Tapiola as a basis of building a bigger urban community. Thus, by the late 1960s the municipality of Espoo felt that Tapiola should be developed further into a regional centre (Tuomi, 1992: 51). From this time what we have seen is the emergence of aggressive pro-growth strategies on the part of the City of Espoo and also Vantaa which have had 'a systematic strategy for growth, and have been active in supplying possibilities for both residential and commercial development' (Laakso and Keinanen, 1995: 125).

> In the beginning Asuntosäätiö and the city of Helsinki had the idea that they take it away from Espoo and Lepavarra had the same ideas and the politicians realised that if the best pieces were going to Helsinki they had lost something. That's why they decided to create the centre of Espoo... and think of it with several centres. [Interview E5]

This characteristic pattern of development in the municipality has also helped shape the startling comparison between Espoo and California made at the outset of this chapter. To recount the views of an interviewee at the Chamber of Commerce, 'this is like California. You need two cars. If you don't have two cars you are in trouble... because we don't have a good transport system. The city centres are not so tightly connected' [Interview E8].

To understand this comparison we need to imagine the Tapiola of the late 1950s or early 1960s. Bus services to and from Helsinki to most parts of Espoo are today extremely good by most European stand-ards. Those between Espoo's five centres are poorer. Indeed it has been remarked that it is quicker to travel into Helsinki and back out than to try to travel by public transport between them. Nevertheless the provision of subsidised public transport has improved enormously and to an extent that few would recognise a comparison with American car-based culture. However, Tapiola and the earliest developments in Espoo stood alone at the time of their initial development and were largely unconnected to Helsinki in any meaningful way by public transport. Moreover, for a time until the formation of YTV (the metropolitan planning body responsible for transport and waste disposal coordination), bus services in Espoo and between Espoo and Helsinki were extremely fragmented.[2] Moreover, it is only more recently still – perhaps in the period 2001–2005 – that the density of population in parts of Espoo has reached levels able to justify today's comparatively frequent services [Interview E9].

As we noted earlier, a proposed metro or light rail link between Espoo and Helsinki had been envisaged since the founding of Tapiola. The proposals over the metro line highlight contradictions over growth versus conservation apparent among different constituencies in Espoo itself as well as generalised cultural differences between residents in Helsinki and Espoo. So, for example, surveys of public attitudes to the rail link conducted in Espoo indicate that residents in the rural north of Espoo are in favour of it presumably because it will have minimal impact upon their environs. However, residents in the urban south of Espoo whom the extension should benefit are on balance against its development, presumably because of what it entails regarding the further intensified development in areas which originally appealed to those moving out of Helsinki. Among Espoo residents as a whole there is also a view that the extension was undesirable because of its presenting a possible dilution of the exclusivity of residential areas.

Attitudes towards public transport have grown to be very different in the two neighbouring municipalities. The proposed rail extension from Helsinki to Espoo would undoubtedly benefit major companies such as Nokia, Kone, Radiolijne and Fortum. These and other companies tend to be located at major intersections along Ring Roads I and II (Figures 7.4 and 7.5). However, they appear to have been largely silent on the subject of the metro extension [Interviews E8, E10]. Indeed, according to one interviewee, 'actually when we are talking about Nokia, the metro is not the question. The question is how fast can the taxi get from the airport to Nokia. And the next question is how can the employees get from

Figure 7.4 The Keilalahti area

Figure 7.5 Offices alongside Ring Road II

their homes to the office. And as we know they are living in their one family houses and then it's more a question of how is the traffic going, not the question about public transport' [Interview E10]. Thus, alluding to the attachment to the private car felt among Espoo residents, another interviewee went on to note how 'It's been said some times, partly as a joke, that people in Espoo don't like the metro because there is no first class' [Interview E1].

In important respects the metro line has been a source of considerable rivalries between municipalities of Helsinki and Espoo and the different micro-models of urban development and social welfare that they embody. The extension of some rail connection to Espoo has been a high priority for Helsinki City but the chief priority in Espoo has been the completion of Ring Road II. Things have come to a head recently with Helsinki-based political representation managing to elevate the extension of the rail line to Espoo within the metropolitan planning agendas [Interviews E2, E3]. Moreover, Helsinki City, which owns a small parcel of land through which Ring Road II will pass, has effectively been blocking its completion using this as a bargaining counter with which to force Espoo City to accept a rail connection [Interview E7].

7.4   A Finnish growth machine?

In all of this a picture emerges of a marked contrast between Espoo on the one hand and Helsinki on the other, in the form of not only urban development but also the urban politics driving this development. In one widely quoted analogy, Vantaa is referred to as the 'poor man', Espoo the 'rich man' and Helsinki being somewhere in between [Interview E3]. This analogy also hints at something of the rivalry that exists between the municipalities. Such competitive relations among municipalities within Finland as a whole, and especially those in the capital city-region, date back to before a tier of regional government was established in the 1960s.

Before this time the competitive pressures to attract a tax base in the form of residents or jobs were apparent in the relations between many municipalities (Sundman, 1991: 92). As such, 'relatively independent municipalities within the region compete with each other by offering the market building sites and projects, which the planning system as a whole could neither effectively oppose nor satisfy. As a result, the urban sprawl has the potential to accelerate . . . furthermore there is no sign of any growing willingness to improve regional co-operation' (Laakso and Keinanen, 1995: 136).

Instead, as one interviewee surmised, there are important differences in the outlook of the various municipalities within the greater Helsinki region.

> They [Espoo and Vantaa] understand the needs of big companies and developers better than Helsinki . . . I think that the background is in the urban history of the three cities. Helsinki has a much longer urban history and it has a much smaller space in which to organise city functions. They have to plan it quite strictly. Espoo and Vantaa have more space and more alternatives. They are in big need of getting especially jobs but new inhabitants too and they are much more liberal or market oriented because they want to have those companies and jobs. [Interview E3]

In recent years this has been manifest in the very visible form of flagship developments for major enterprises attracted to Espoo, often from just across the water in Helsinki. Companies such as Kone, Nokia and Siemens have all been attracted to Espoo, less by financial incentives than by the availability of land, less restrictive planning approaches and labour-market conditions after having been refused planning permission for new or expanded office blocks within Helsinki. The most distinctive of these office strips has burgeoned rapidly in the Keilaniemi–Keilalahti area. The headquarters of Finnish company Fortum stood here alone among trees from the mid-1970s but has been joined recently in the space of the period from 1995 to 2005 by others such as Nokia, Kone and Radiolinja to create an imposing cluster of brand names facing back towards Helsinki.

The municipality's disengagement from the development process that at several junctures saw calls for Tapiola and Leppävaara to be incorporated into Helsinki actually stems from several interrelated factors. First, and as already noted, at this early stage of urbanisation the municipality certainly lacked the planning regulatory powers and the fiscal base to underwrite extensive service and infrastructure provision to new housing developments.

Second, and related to this, is that fact that the human and technical capabilities of the municipality have been underdeveloped in comparison to the planning department of Helsinki City which remains very well resourced in comparison to Espoo and other large municipalities. At the outset of serious development of Espoo there were just two planners employed by the municipality. One interviewee recounted the extreme case of the 'planning' of the Livisniemi area in Espoo. After a developer had acquired land in the Livisniemi area it pressured the city council to grant permission and produce a detailed plan for the area. The two employees were told to make a detailed plan for the area within one week. They took an old plan from Helsinki City Council Planning Department. This plan was imposed upon the area without regard for nature or topography. It remains a detailed plan that is said to be valid today [Interview E7].

Third, municipalities' statutory powers in the making of agreements with developers were limited at this time and were only extended in more recent legislation. According to several interviewees the bargaining position of municipalities has increased and the sorts of agreements characteristic in the growth of Espoo would not happen today [Interviews E4, E6, E7]. Many opportunities have been open to developers and construction companies given the lack of agreed strategic plans for much of Espoo. Over the years there have been

> many, many proposals or master plans but the political decision-making has always been very hard. So that it's a joke that it takes 20–30 years to get a master plan decided in the city of Espoo. And even now, in the northern part of Espoo which is sparsely populated they have a master plan but in the southern part of Espoo, which is the main part of jobs and population, they don't have any approved plan. They have all kinds of proposals during 30 years but not a decision. [Interview E3]

Thus, in the early days there was a 'communal politics of the "wild west"' in which there were 'unofficial negotiations between developers and an inner circle [of councillors] who would make deals very freely without control' [Interview E4]. Indeed, the procrastination over such plans has itself been suggestive of the close articulation of political and development interests according to one interviewee. 'I have got the feeling that it was also in a way a political failure. They didn't want to accept these goals . . . Or they haven't actually decided what the main goals are. It was just up to the planners to come and have some ideas of how it should be, and then it was so easy to throw it back and say we didn't want it' [Interview E6].

Fourth, over and above these factors has been what one might describe as an enduring legacy of the history of development process in Espoo which in turn has created a culture of working in the municipality itself. The rural conservative politics of Espoo has mutated into a harder edged pro-growth politics in which elected local politicians and, until recently, officers have contrived in this spontaneous pattern of development. This conservative politics resonates historically with Espoo's geographical positioning.

> This area, Espoo, especially Tapiola, has been western oriented over time and it can be seen in the political life of the city still today because the conservative party [the National Coalition Party] dominates here – which is not the fact in most other bigger cities in Finland. So this very strong right wing attitude is typical of this city. [Interview E1][3]

The relative disengagement of the municipality from shaping urban development may reflect a more general characteristic of Finnish local government. Mauriten and Svara (2002, cited in Goldsmith and Larsen, 2004: 130) characterise Finnish mayors as caretakers with consequent weak leadership in both public profile and policy leadership. Certainly it was the view of the interviewees that only with the recent arrival of Mayor Marketta in Espoo has there been a concerted effort to eliminate vestiges of development practices long tolerated and enduring in Espoo. As an interviewee noted,

> It was quite common in the 1960s and still in the 1980s that the chairperson of the city planning boards were from the construction sector or something but it's not so common any more. But in Espoo we now have the chairperson of city planning board is a real estate businessmen. So it's the only example nowadays. [Interview E3]

The sorts of patterns of business political activity within the local and city-regional context in Espoo and Helsinki city-region are schematised in Figure 7.6. What we see here is a pattern of urban politics that comes close to Molotch's concept of an urban growth machine. Undoubtedly large companies such as Nokia and Kone wield some influence locally as they do nationally, though Castells and Himanen (2002) argue that national policy has not been geared to the special interests of firms such as Nokia but to wider industry needs. In similar vein, interviewees suggested that the largest companies rarely comment publicly and are invariably 'correct' in the communication of any views they may have about major municipality-wide issues, but instead content themselves with more specific real-estate issues such as land for expansion or access [Interviews E11, E12]. Organisations such as the Chamber of Commerce which has had a branch in Espoo since the 1980s are in regular dialogue

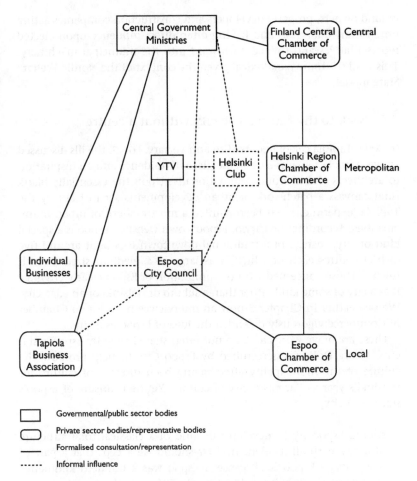

Figure 7.6 Map of business interest representation in Espoo

with Espoo City and play an important consultative role in day-to-day policy development compared to the largely marginalised collection of smaller businesses represented by the Tapiola Business Association [Interviews E8, E11, E14].

The key business interests that influence municipal agendas have been and continue to be land based. Organised broad-based manufacturing and retail business interests have been formally incorporated into municipal politics more recently but continue to have little significant impact. Of far greater significance as we have seen already has been the role of land owners, property developers and construction companies acting on an individual basis and having often direct influence upon elected representatives and thence downward into the municipal machinery. This is all the more paradoxical given the context of the Nordic Welfare State model.

7.5   Back to the future? The city without a centre

In 2003, Tapiola celebrated its 50th anniversary. For all the ills discussed above, Tapiola remains a desirable place of residence and an inspiration to architects and planners internationally. Upon the essentially blank rural canvass a patchwork of urban developments set in train by the Tapiola 'experiment' has been painted a most modern of urban muni-cipalities. According the City of Espoo's own website, 'Espoo is a special kind of city. Instead of a traditional downtown it is built around five district centres each of which is as large as a medium sized Finnish town'.[4] Those concerned with the promotion of Espoo clearly envisage it as a city of some kind rather than a suburb of Helsinki or an edge city. We saw earlier in Chapter 3 how an interviewee from Espoo Chamber of Commerce vigorously defended the idea of Espoo as a city.

These are brave words and are not often shared even by Espoo resid-ents. Georg Dolivo was recruited by Espoo City to help promote the culture of the municipality after having been director of culture for Helsinki's year as European city of culture. Yet he is unsure of Espoo's status as a city,

> should Espoo try to develop this kind of a classical total structure of a city with all its elements? Or should we accept that a part of it is outside Espoo? . . . Because if Espoo was a hunderd kilometres away from Helsinki it would be totally different. There are two ques-tions that arise from this? One is what should Espoo do? The other one is how should we relate ourselves financially and technically to the things that are happening in different municipalities? And this second question, I feel, is very delicate politically. [Interview E13]

And as others less attached to any civic promotional role for Espoo are able to observe,

> Espoo itself is not a city. It's a combination of different districts that have been built up independently from one another. So in possibly a hundred and fifty years from now it will then come together and

it will be considered a city but by that time it might already be incorporated with Helsinki. It's a big political question. [Interview E2]

Espoonkeskus may be the administrative centre of the municipality and may also be home to Espoo's oldest building, a twelfth-century church, but it lacks any of the other functions and just as significantly the vibrancy of a city centre. Tapiola has the vibrancy, some of the retail and cultural amenities of a city proper but not the administrative functions. Each of the five district centres arguably lack anything that could act as a symbolic space within which civic identity can be expressed. Where, then, is the centre of Espoo? One could argue, somewhat facetiously, that the 'symbolic centre' of Espoo is the Länsi-auto Stadium – the ice hockey stadium in Tapiola. This huge building and its associated parking lots meld national and local culture. In function the stage for a national sporting obsession, in name it invokes the local attachment of Espoons to their cars (Figure 7.7).[5]

If Tapiola garden city was a vision of the future – a better future – for Finns, Espoo is again at the forefront of thinking regarding the future of urban form and function. Espoo is to be the site for a new planned settlement of the future – orchestrated this time by the council and also with significant input from Nokia. A site in the undeveloped Suurpelto area

Figure 7.7 The Länsiautostadium

of central Espoo will be the location for a new planned settlement. The settlement is part of an EU-wide network of similar urban developments in Barcelona and Amsterdam and the famous Sofia Antipolis in Southern France (Anteroinen, 2003: 50). Ideas for the planning of this new settlement have apparently also been gleaned from the Netherlands, the US, UK and Singapore (Tornroos, 2004).

For Nokia's part it was described how 'we want to participate for this planning process in order to create a good appropriate platform for future society... There is a vision... that work, housing, free time etc. will get as large a diversity as possible. Now what we are doing is to clarify what it could be but at the end of the day, how this diversity will be used totally depends on the future land owners, companies or developers etc. but the platform will be there.... This is not going to be Tapiola but it will be something else that will be remembered' [Interview E12]. The expectation is that the likes of media cafes, libraries and sports venues could be micro-planned in such a way as to create in miniature the sorts of networking that Castells and Himanen (2002) suggest is at the root of Finland's successful information society. Thus, 'In this new urban setting we'd also like to see how the physical environment and town planning can promote "accidental" encounters and hence communication and creativity' (Kokkonen, quoted in Tornroos, 2004: 6).

In 2003, the 50th anniversary of Tapiola garden city, the parallels between the two developments were indeed striking. The 'the Suurpelto project has similarities with Tapiola in the sense that the land area in question is large, integrated with nature, a significant part of southern Espoo, and, again we are striving to build a truly new community' (quoted in Anteroinen, 2003: 51). In this respect we have come full circle. The Suurpelto development probably represents the last such major integrated district to be planned and built in Espoo for the foreseeable future. If Tapiola represented the start of a process that paradoxically produced an American-style landscape within the Nordic welfare state model, the Suurpelto development provides the other bookend to Espoo's story of rapid urban growth.

7.6   Conclusion

The comparison drawn between California and Finland's second largest city may at first sight seem highly fanciful. However, one purpose of this chapter has been to demonstrate that, putting to one side the specificities of edge city forms found in the US, in terms of the typical of processes of development and functioning, this particular European post-suburban settlement approximates quite closely to a US outer suburb or edge city. Even in terms of the physical form, Espoo, in some instances, comes quite close to the stereotypical edge city low-density residential, office and retail environment. Its urban landscape is littered with symbolic references to the US culture, and residents and business people self-consciously identify themselves with such symbols in contrast to their neighbours in Helsinki. The remarkable thing is that the seeds of this diffuse, centreless, urban form were sewn in the sorts of garden city planning ideals that were intended to produce a better city form and environment. And in this respect too, the story of Espoo is one that resonates with the planned sprawl of the US.

The most striking parallel to the sorts of suburban and post-suburban developments found in the US is to be found in relation to the underlying *process* of urban development itself and the sorts of politics surrounding it. Unlikely as it seems, Espoo's urban development certainly bears most of the hallmarks of Molotch's (1976) growth machine. The urban politics of Espoo has for the last four decades centred on the desire for economic growth and has been fuelled primarily by land-based business interests such as land owners, construction companies and developers, rather than other major companies with a lesser local dependence.

8
The Croydonisation of South London?

Croydon has become a creature of the depths, a subtopian city-state; constantly reaching out to devour the lesser hilltop developments of South London.

Iain Sinclair, *London Orbital*

8.1 Introduction

Well over one hundred years before Iain Sinclair detected his creature of the depths, commentators had been aware, it seems, of the 'Croydonisation of South London'. By the 1890s and with the new London County Council barely installed, the general trend of population growth in the outer rings of Britain's major cities including London could be observed. In 1891 the implications for London already seemed clear to Low (quoted in Young and Garside, 1982: 107): 'It will be a London of suburbs. . . . Not one but a dozen Croydons will form a circle of detached forts around the central stronghold' with the people of London dwelling in 'the depths of the Home Counties'.

Croydon is by far the largest of the edge city municipalities in the network and has the largest single concentration of employment (Figure 8.1). Its miniature Manhattan skyline is home to an office and retail complex broadly the size of free-standing cities such as Liverpool and Newcastle except that it is part of the continuous urban fabric that is greater London. At the beginnings of the 1800s Croydon had a population of a little under 6000. During the 1800s Croydon further established itself as a sizeable market town but its population grew rapidly in the late 1800s and early 1900s as it became a dormitory

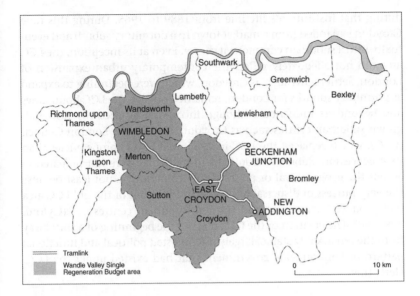

Figure 8.1 Map of Croydon

suburb for London. Today, Croydon boasts a population of 335,000 and a sizeable service-sector economy.

Our case-study post-suburban settlements each highlight different aspects of the eccentric geometry of post-suburban areas and their centrality to contemporary urbanisation. In Noisy-le-Grand the vacuum of local identity was a product of the multiple and overlapping non-local state structures in which it was entangled. The edge entrepreneurialism of Getafe rests, significantly, on its geo-political manoeuvring within the metropolitan institutional and political setting. Croydon is large enough as a post-suburban *place* to be both internally fragmented on the one hand and a platform for entrepreneurial local institutions to have enlarged spheres of influence within emergent South London administrative *spaces* on the other. In its metamorphosing as a place and in the entrepreneurialism of its local institutions, notably the council, we can see what Ian Sinclair refers to as Croydon's 'devouring the lesser hilltop developments of South London' (Sinclair, 2002: 323).

8.2 Contrary Croydon

As a County Borough, Croydon remained outside and independent of London-wide government under the London County Council (LCC) during that institution's lifetime from 1889 to 1965. During this time Croydon expanded from a market town to a dormitory suburb and eventually into a suburban commercial centre. Even at its inception, the LCC area did not adequately reflect the contemporary urban expansion of London. Settlements such as Croydon, which were beginning to expand as dormitory suburbs for London, remained outside the LCC boundaries and beyond its influence. Although incorporating these by now fullgrown suburban centres such as Croydon, the Greater London Council (GLC), which replaced the LCC in 1965, had fewer competencies than its predecessor. 'Suburban boroughs in particular were not prepared to accept the geographical or political subordination... not least because the very process of dispersal of people and jobs out of the old LCC area had produced some new and thriving suburban centres...' (Gyford, 1994: 80). The creation of the GLC marked the beginning of a shift away from the remarkably stable, largely uncontested political and functional pattern of London-wide government that had existed under the LCC. Here, then,

> If the boroughs... made active incursions into strategic planning issues during the GLC era, they also of course focused very firmly in planning their own territory. As well as promoting the idea of a polycentric Greater London, they proved equally enthusiastic about imparting a monocentric character to their own individual boroughs. (Gyford, 1994: 80–81)

Partly a result of this and of the continued strength of suburbs in the UK when compared to those in, for example, the US and France, London has recently been depicted as a 'city of villages' (GLA, 2002). Croydon provides a prime example of the sorts of suburban independence described by Gyford. Croydon council was significantly opposed not just to the GLC but also to central government (Saunders, 1983). And as an interviewee described 'it's always been hugely independent. It's always seen itself as separate. It's a cultural thing as much as anything else. It's certainly a culture within the council' [Interview C1]. In this period in which the GLC superseded the LCC, Croydon was undergoing dramatic expansion into a significant suburban commercial centre in its own right. The laissez-faire form of office development which established Croydon's familiar, if much derided, office landscape was nevertheless facilitated by a determined piece of opportunism or municipal entrepreneurialism on the part of the then Conservative-controlled Croydon Council (Phelps, 1998; Saunders, 1983). This propelled Croydon, at least partially, into an edge city in *function* (though not in form) at around the same time that North American edge cities began to emerge in earnest. It is during this period and with subsequent growth that Croydon metamorphosed from an exporter of labour to a major employment centre in its own right. Partly as a consequence it has also emerged as the most self-contained (measured in terms of the proportion of the workforce that both live and work within the borough) of London boroughs (Croydon Partnership, 1998: 11).

From the dismantling of the GLC in 1986 to the formation of the Greater London Authority (GLA) in 2000, London-wide government became more fragmented under a profusion of quangos, partnership arrangements and service delivery organisations with overlapping territorial jurisdictions (Gyford, 1994; Newman and Thornley, 1997). Despite strong divisions between London boroughs along party political lines, the polycentric system of metropolitan government which emerged under the GLC was held together by a certain pragmatic mode of joint working (Hebbert, 1992). The new London-wide institutions – the GLA and its development agency arm, the London Development Agency (LDA) – have retained and refined some of the neoliberal fragments of pan-London coordination established since the dismantling of the GLC. Under the GLA then,

> the neoliberal emphasis on . . . competition and innovation, and the recognition that London's economic growth and prosperity is strongly linked into processes of globalisation, have produced the embrace of a pro-growth, pro-business globalisation agenda that reflects a major shift in attitude from the GLC's 'restructuring for labour' agenda of the mid 1980s. (Syrett and Baldock, 2003: 81)

Thus whilst London business interests proved effective in shaping policy agendas prior to the creation of the GLA, they did so at a national, central government level but have since come to be an integral part of an emerging London urban regime (Thornley *et al.*, 2005: 1964).

The GLA's recently published *London Plan* can be distinguished from two previous plans for the capital by virtue of the high degree of political steering in its formulation and the emphasis upon retaining the competitive position of London as a world city. Business has been particularly effective in shaping this agenda with the GLA evincing a degree of strategic selectivity (Jessop, 1990) in favour of business in comparison to other stakeholder groups including London boroughs (Thornley *et al.*, 2005).

The idea of polycentric development has been retained and enhanced under the workings of the GLA and LDA and in the recently published *London Plan*. Thus the *London Plan* explains that 'the Mayor supports polycentric development across the mega-city region in which central London, London's town centres and the towns in... other regions develop in a complementary manner' (GLA, 2004: 18). Yet the GLA and LDA have also made a more concerted attempt to plan and integrate activities on a sub-regional basis, albeit with an emphasis upon the East London Thames Gateway. These sub-regional boundaries have themselves been an enduring issue within the greater London institutional scene.

> the issue there's always been with London is boundaries, because these are artificial boundaries and they never work properly for every activity. They might work for training but not work for transport. They might work for patterns of employment or unemployment but it might not work for, you know, other activities like regeneration. [Interview C2]

Taking the Learning and Skills Council sub-regional boundaries, the new *London Plan*, published in 2004, and the work of the GLA and LDA have effectively established a new set of sub-regional territories for the purposes of spatial planning and the delivery of some services. Some of the implicit characterisations of the different London sub-regions became apparent in the draft Plan's examination in public (EIP). In this regard, for example, South London boroughs and other representations argued that the growth potential of the sub-region had been seriously underestimated (Richardson and Simpson, 2003: 57). As one interviewee described, the draft plan

> basically concentrates on central and East London Thames Gateway... And there was a feeling that north West and South London were slightly underplayed. A feeling we shared to some extent. We thought that the contribution that those areas could make, particularly in terms of jobs but also to an extent housing, had been a bit underplayed by the plan and we wanted a bit more emphasis on that... including the role of town centres in those areas, which we thought hadn't been sufficiently developed in the original draft plan. [Interview C3]

Although Croydon, for instance, is recognised as one of two strategic office locations outside of Central London, the *London Plan* notes that development opportunities are generally smaller in scale in South London when compared to the other sub-regions (GLA, 2004: 276 and 275 respectively). Moreover, according to the views canvassed, 'the relatively low job growth forecasts reflected what many saw to be a tendency in the draft Plan to "write off" South London as a dormitory area for the major growth in the Central and East London sub-regions' (Richardson and Simpson, 2003: 57).

Certainly, for several outer London boroughs including Croydon, the new GLA and LDA environment embodies a continuity from previous eras of pan-London governance under the LCC and the GLC in that 'for the outer London boroughs I think there is a feeling that they are only half involved in the London agenda and they are half elsewhere' [Interview C3]. Nevertheless, this enduring problematic has also, as we will see, represented an opportunity for a borough like Croydon to exert a wider influence in South London and beyond. Despite its long being a post-suburban *place*, as signified by powerful senses of independence from central and greater London, Croydon is also internally fragmented. Moreover, in the recent more fragmented era of London-wide governmental arrangements, Croydon emerged as a leading exponent of partnership working through which there has been a partial 'Croydonisation' of South London as Croydon-based institutions have extended their sphere of influence laterally to neighbouring South London boroughs.

8.3 Croydon's urban regime

There is a long established history to the pro-business politics of Croydon Council which Saunders (1983) traces back to the 1800s. More significantly the entrepreneurialism of the Council was revealed in the economic boom years of 1950s and 1960s Britain. The main ideas underlying the central Croydon of today were laid out in a plan prepared by the Reconstruction Committee in 1945 and enshrined in the Development Plan of 1951. The Council sought and obtained wide-ranging powers of compulsory purchase under the Croydon Corporation Act of 1956. Its newly gained powers coincided with the ending of building restrictions in 1954 and an accompanying building boom and concerted efforts to decentralise office functions out of central London. Fully 20 per cent of all offices and 30 per cent of associated jobs relocating out of central London in the period 1963–1973 went to Croydon (Harris,

1993). As Marriott observed, 'the most sensational phenomenon thrown up by the office boom in South East England was the development of Croydon.... There has been much talk of... decentralised office centres but Croydon is the only centre worthy of the name' (Marriott, 1967: 185). Having obtained these powers the Council was content to let market forces dictate, and further facilitate, the pace and style of development. The resulting boom in development was not without its critics. Anti-development pamphlets complained how 'Croydon Council aided and abetted private development, and gave private enterprises a free hand to operate unhindered.... A "living" town has been destroyed and replaced with a tombstone to commerce' (Suburban Press, 1972 reproduced in Harris, 1993).

Offices and shops were constructed rapidly with little regard for architectural design quality (Figure 8.2). Yet by the end of the initial office boom the council celebrated the redevelopment

> the new and impressive town centre will form... a truly contemporary and quite splendid focal point for this great London Borough. Vast office blocks have made a new skyline, their towering elevations dwarfing older buildings... enough has already been accomplished

Figure 8.2   The Central Croydon office complex

to show that this is truly becoming a city of the present and also of the future. (London Borough of Croydon, 1969: 43)

During the 1970s, 1980s and into the 1990s a growing emphasis upon strengthening design standards of buildings in the central area and improving the attractiveness and interest of the central area of Croydon was to be found in local plans (Phelps, 1998). As one interviewee described,

> Croydon was not looking outwardly enough at that point – it thought it would survive purely as a strong office market. But we've come to realise that it's a lot more than just a good office base. You've got to have all the support and all the supporting businesses. I think that was a hard pill to swallow. And of course the image kicked in because those blocks became tired looking... We're still in the cycle if you like of trying to get over that bad image, that legacy from the 60s. [Interview C4]

By the 1990s, then, any complacency over the continued standing of Croydon's central office complex had vanished, as a skyline that had been considered 'the most consistently modern in Britain' was recognised as part of the problem of attracting further office and retail development. As one local history commented, 'there is little relief to the eye from the harshness of many of the buildings; routine commercial architecture has, in the main, been the order of the day' (Gent, 1988: 59). One of the major concerns of the council and of its marketing arm Croydon Marketing and development (CMD) has been the image of Croydon [Interviews C5, C19]. As one interviewee described,

> One of Croydon's biggest challenges as a town is perception management. Croydon is perceived as a 1960s concrete jungle.... And when ever you read the press it's always 'who'd be in Croydon? What an awful place to be'. It's always talked of negatively as a town. [Interview C6]

The municipal entrepreneurialism forged in the 1960s has provided a lasting legacy. However, in contrast to the sweeping and wholesale changes made possible in the 1960s, the Council's abilities to reshape and renew central Croydon are today more constrained than in the past. After a period in the mid-1990s when there was somewhat of a downturn in the local office market and a widespread questioning over the future of Croydon as a major employment centre, the Council mounted a 'Croydon – the Future' exhibition in conjunction with the Architecture Foundation. This was aimed at developing innovative ideas for the redevelopment and re-imaging of Croydon. In the wake of the exhibition the Council along with Eckbo, Dean and Williams (EDAW) consultants began to draw up a new Master plan vision for central Croydon entitled *Croydon 2020* (Croydon Council, 1998). The context for this plan is more complicated in that the Council is faced with having to try to generate consensus regarding development priorities across a swathe of territory, the majority of which it does not own and over which there are ownership and lease conflicts [Interview C4]. It has had to be entrepreneurial in a way different to that in the 1950s and 1960s – forging a broad-ranging public–private partnership of central Croydon stakeholders with private investors in particular being sought to provide the bulk of a total £600m of investment estimated in 1998 to be needed for the envisaged redevelopment (Croydon Council, 1998: 3).

From at least the 1950s and 1960s and into the 1990s and the *Croydon 2020* vision, it is clear that the Council has been central to the facilitation of the redevelopment process in central Croydon. In this respect,

> Croydon is very proactive as a borough. It does appear to take control of the development process and push its own things far more than other boroughs I can think of and far more obviously than you could say about individual private sector developers. [Interview C1]

Yet the Council's prominent role in orchestrating the development process and the various actors has been a source of some conflict within the borough and within the greater London institutional setting.

Although, as we have seen, South London as a whole was underrepresented in the *London Plan*, Croydon town centre is designated as the only 'opportunity area' within the sub-region. Yet, 'LB Croydon were disappointed that the draft Plan in their view underplays the strategic importance of Croydon' (Richardson and Simpson, 2003: 58). Moreover,

> Croydon is the one 'opportunity area' in the London Plan. . . . Having said that, that's something of a mixed blessing for Croydon. They've got their Croydon 2020 which covers the same area and I think there is going to be some care about how those two documents fit together. [Interview C1]

As Richardson and Simpson (2003: 58) noted, *Croydon 2020* was more ambitious than the policies contained in the *London Plan*. Another interviewee was more explicit about the potential for conflict between local and metropolitan priorities:

> whatever anyone says, we are not really in London and therefore when Ken decides to get involved with us things tend to take a little bit longer because he actually doesn't perceive us as being part of London... And because Croydon therefore is not at the heart of London, there is always a bit of a difference of opinion as to what the priorities are there. [Interview C6]

Although considerable effort was taken to consult and include all the relevant stakeholders in the *Croydon 2020* 'vision', the fragmented landownership pattern coupled with the interests of private-sector developers have generated conflicts. As one interviewee noted, 'I'm terrified that Vision 2020 is fine in theory but this is the practice. People are arguing, putting in counter planning applications and all the rest' [Interview C7]. Two notable proposed developments have crystallised some of the conflicting interests between the Council and developers and between the council and the GLA.

First, then, there has been the conflict between the GLA and Croydon over the Park Place development in central Croydon which involved a major redevelopment of existing retail and office space by Minerva. Interestingly, this seems to represent something of a battle over old and new landed and commercial interests within the Croydon sphere. In the Croydon of the 1960s and 1970s, the Whitgift Foundation was a major land owner and came to have significant commercial interests with the development of the Whitgift shopping centre (Saunders, 1983). In recent years the Foundation largely has played a passive role in further developments [Interview C1] except where its established interests have been affected. An additional shopping centre – Croydon Centrale – was recently opened. The Whitgift Foundation's established retail interests have been more directly confronted when Minerva began to buy-up and parcel together land in order to create a single major new retail scheme in central Croydon. The Whitgift Foundation having interests in the existing major shopping scheme in Croydon and being one of several land owners resisted purchase of a small piece of land in the light of the competition to its own retail development represented by the new Minerva Park Place scheme. The GLA also had reservations about the design quality of the scheme which they believed were not stringent enough for such a landmark project [Interview C1]. Curiously, despite increased emphasis in the design standards apparent in successive local plan documents (Phelps, 1998), design quality appears to have been a constant issue in the appearance of Croydon.

A second proposed development over which Croydon has attempted to assert its independence from metropolitan and central government is the redevelopment of a major 'gateway site' close to East Croydon station. This particular development is a prime example of how the council is

> quite a wayward borough. It's got its own agendas which they are
> wedded to which aren't necessarily GLA's agendas. They are wedded
> to particular projects that they want to see through.... They quite
> usually reflect the quite long-term almost personal political agendas
> of the borough. [Interview C1]

Croydon 2020 earmarked a major central site for an arena and associated retail and housing and office elements as *the* flagship development. Whilst Croydon remains a regionally significant commercial and retail centre, the same could not be said of its cultural amenities. In particular the arena project is important to Croydon in order to renew such a regional role in the light of the fading appeal and facilities offered by the existing Fairfield Hall built in the 1960s [Interview C8]. The council has given its backing to the arena scheme of Arrowcroft developers. The GLA remains sceptical of the viability of the scheme. Moreover, the land owners, Stanhope, have their own plans for the site which do not include an arena. Croydon council has indicated that it will grant planning permission for its favoured Arrowcroft scheme and has threatened to use powers of compulsory purchase to ensure its development. Stanhope meanwhile has argued that it has popular support among residents of Croydon and the issue is set to be resolved by London Mayor Ken Livingstone (Davey, 2004).

What these two examples highlight is the determination of the Council to assert its own independence within broader institutional arena and orchestrate the development process in a way which is quite unusual in the UK. Marriott (1967: 184) likened the autocratic leadership provided in Croydon Council by Sir James Marshall throughout the 1950s and 1960s office boom to that of an 'American-style town boss'. Since this time an entrepreneurial stance has become more diffused and firmly embedded throughout the council machinery and other local institutions. As such, one study has suggested that 'Croydon begins to approximate a US style of urban regime built around local economic development issues, bipartisanship and close public and private sector relations and partnerships' (Dowding *et al.*, 1999: 515–545).

8.4   The Croydonisation of South London

The *London Plan* notes how the South London sub-region 'has strong radial as well as orbital linkages to the other sub-regions' (GLA, 2004: 274). Indeed, the emphasis on sub-regions in the London Plan appears to be a response to the fact that 'many boroughs already look wider than their own boundaries, plan with their neighbours and work with the many institutions now operating at a sub-regional level' and a recognition of the permeability of these regions (GLA, 2004: 221). Croydon as the most populous of London boroughs seems to embody these linkages. It stretches radially from inner-city-like wards in the north to the dominant commercial complex of central Croydon and out to the relative affluence of what Saunders (1983) refers to as the 'deep South' (Figure 8.3) (which borders onto the stockbroker belt of Surrey). Yet again there are contrasts laterally from east to west in the borough. The New Addington area at the eastern edge of the borough is essentially two

Figure 8.3   The 'deep south' of low-density housing, Riddlesdown

Figure 8.4   New Addington housing estate

peripheral housing estates (one a philanthropic estate built in the 1920s, the other a council estate built in the 1970s) somewhat detached from the rest of Croydon (Figure 8.4). Purley Way, on the western edge of the borough – referred to as 'shed city' by a senior economic development officer [Interview C9] – has something of the spontaneity and free-market character of an edge city (Phelps, 1998). Significantly, this is the site of London's first commercial airport which was quickly overtaken by the development of Heathrow and consequently was redeveloped into an industrial estate.

As such Croydon is geographically fragmented and displays only partial elements of the North American edge city model as one inter-viewee identified.

There is a gap between Croydon as a centre and Croydon as a borough. Because Croydon as a borough is huge.... as a centre, Croydon is clearly an office based retail employment centre. It's got good public transport links but it's still got very much a car based culture which comes across from the 1960s and 1970s. So it does have those elements about it. But as a borough it's too big and too differentiated to fit into those easy American [edge city] definitions. [Interview C1]

It is clear from casual observation that Croydon is not a single self-contained place with a single identity. An article celebrating Croydon's rise to prominence in the 1960s made note of 'a splendid new flyover which does not appear to go anywhere at the moment but eventually will fit neatly into the new southern motorway system' (Hodson, 1971). There is an irony here, for whilst in many respects Croydon does enjoy enviable communications within a South London and south-east England setting, adequate road links to the M25 to the south remain the one main problem of accessibility for the borough.

Croydon's flyover still does not go anywhere. Yet what Croydon's radial relationship to central London and the lateral Croydonisation of South London boroughs reveal are Croydon's character as a post-suburban gateway. This is stressed explicitly in the recent city status bid and in the form of 'Croydon Gateway' – one of the largest development sites in the south-east of England, which plays a prominent part in the redevelopment of central Croydon envisioned under the *Croydon 2020* master plan (Croydon Council, 1998).

Today, this gateway property of Croydon underpins its central Croydon office-retail complex (Marriott, 1967; Phelps, 1998) that arguably constitutes an edge city development in function. But it also has a longer lineage in other land uses closely associated with post-suburban development. Croydon was once a 'gateway to the planet' – a whimsical reference to London's first airport located in Croydon but which ceased operations in 1939 (Calder, 2000). Indeed Croydon's early prominence as an example of the airspaces that many now regard as quintessential examples of modernist architecture and ingredients of post-suburban development was bemoaned by Le Corbusier who argued that 'a triumph of wings is well worth the loss of a capitol. We have suffered long enough that sordid exile in the suburbs of Bourget or Croydon' (quoted in Pascoe, 2001: 132). The central office-retail complex represented a further partial transformation of the borough to include elements of the function of an edge city but was overlain as it were upon previous market town and dormitory suburb and airport-industrial functions of the borough. The identity of Croydon is also therefore fragmented temporally as another interviewee highlighted.

> there is no community of Croydon. There is a day community – called the people who work here. There's a night community – the residents. And there are people coming into us, and around us and over us that makes it quite difficult to generate this community spirit. [Interview C7]

Moreover, as a 'city in waiting', Croydon's influence encroaches laterally into neighbouring South London boroughs. Most immediately the
sheer size and gravitational pull of Croydon's office and retail centre
has prompted neighbouring boroughs to differentiate their town centre
shopping areas to avoid head-on competition. This lateral encroachment
manifests itself in a political form. As one interviewee, alluding to a
phrase that had become familiar in Croydon Council circles, suggested,
'Croydon has a policy of being promiscuous where partnerships are
concerned' [Interview C10]. Here, the suggestion was that Croydon
Council had actively sought to engage itself in as many partnerships
as possible. Such partnerships have seen the Council expand what Cox
(1998) would term its 'spaces of engagement' locally – within South
London but also nationally and internationally.

Within the South London sphere, an important space for engagement has begun to open up. At the outset new sub-regions began to
be imposed as a result of rationalisation within the Greater London
institutional sphere. As one interviewee described it,

> along came the Mayor and the LDA and it was very apparent that this
> ridiculous geography that we'd been given, this sort of outer belly
> of, if you like, South London – it wasn't even South London, you
> see there is no identity – was crazy. It wasn't going to have a voice.
> [Interview C7]

However, the relevant institutions have themselves further rationalised
and adapted their territorial jurisdictions to create an increased degree
of cohesiveness. Such cohesiveness has been borne out of necessity as
another interviewee noted.

> The money going East is part of the major inward investment strategy
> for the Thames Gateway area. There's an awful lot of money being
> spent down there. . . . It's not an equal pattern of spend throughout
> London. So it has meant that South London is really coming together
> finally in one voice because we are all hungry. [Interview C11]

Within this context of coalescence in institutional territorial jurisdic-
tions, possibilities for Croydon to take a lead have been recognised
[Interview C2] and, to an extent, seized. This has a longer history which
stems from the relative independence of Croydon from Greater London.
However, if anything, it has become more apparent in recent years.
Thus, as an interviewee explained,

We got the reputation, we are now called 'two jags Croydon' which is a bit unfortunate... So I think they [other South London boroughs] look at us as threatening at times. I think they look at us in history as wanting to take the world over. We in our strategic plan say we want to be the capital of South London. But on the other hand there is much more joined-up working between the six boroughs because they realise the writing is on the wall. [Interview C7]

The evolution of the London Wandle Valley Partnership provides one illustration of the porosity of borough boundaries and identities within South London, and of Croydon Council's active pursuit of larger spaces of engagement for itself. The London Wandle Valley partnership grew out of formal joint working arrangements for a Single Regeneration Budget (SRB) area straddling parts of Wandsworth, Merton, Sutton and Croydon. It has outlived this with informal joint working across an expanded area covering the whole of these four boroughs and beyond in the emerging South London Partnership. As one interviewee described,

Over the years with the London Wandle Valley Partnership we've broken down the initial suspicions and prejudices. I mean we detected quite a lot of animosity towards Croydon. You see Croydon, in South London terms, is quite a significant... local economy... I think there's a feeling that Croydon is pretentious, that the aspirations of Croydon are about forming a greater Croydon. But we play that down. We genuinely believe in working on a South London basis. [Interview C10]

In the local setting of South London, Croydon is highly active in a range of partnerships. Its influence is probably, as one interviewee suggested, greater to the west where Croydon plays a part in the South London Partnership (formerly the Wandle Valley Partnership). Nevertheless Croydon Council and other Croydon-based organisations (such as the Chamber of Commerce) have been active in attempts to integrate activities on a South London–wide basis, with Croydon itself being the preferred location for many of these emergent South London bodies.[1]

These same sensitivities among neighbouring southern boroughs were also thrown into sharp relief with the progress of the South London Tramlink project. As one interviewee from the Council described,

You almost detect a resentment by Merton and others about Tramlink. Because Tramlink was driven by Croydon Council and then

London Transport et al were brought on board. . . . I suppose it's that competition thing . . . it comes back to what I said – the view of greater Croydon. Croydon are pushing this, Croydon are pushing that. We try to play that down . . . [Interview C10]

So objections by Bromley Council to Tramlink and the eventual positioning of its eastern terminus at Beckenham seem testament to a desire to prevent the 'Croydonisation' of Bromley. These sentiments are clearly visible in the representations made by the MP for Beckenham, Piers Merchant:

I think it is very important for me to stress that though people looking at a map might say this is south London and Croydon is quite near Beckenham, there is a very strong historical and natural divide between what effectively was Kent and what effectively was Surrey . . . So they are crossing a border, which might not appear on a map or in London Transport plans, but in terms of perception, it is very important for people in the area. (Merchant, 1994)

Tramlink became operational in 1999 and runs from Wimbledon in the west to Beckenham in the east (Figures 8.1 and 8.5). In 1990 the Council jointly promoted the Tramlink project with London Transport embarking on a massive consultation exercise and winning over 80 per cent of respondents from the Croydon population to the virtues of a light rail system. Echoing the 1960s when a private Parliamentary Bill paved the way for the redevelopment of central Croydon, the Croydon Tramlink Act received Royal assent in 1994. Tramlink proved an early model of New Labour's preference for Private Finance Initiatives (PFIs) but crucially was the last transport project permissible under the Parliamentary Bill procedure and benefited, as a result, from a more generous central government allocation of funds than has been available in subsequent PFIs. Overall it does appear that Tramlink has benefited Croydon within the South London context.

We've been very lucky. It's called Tramlink but everybody calls it Croydon Tramlink. It's associated with Croydon and I think Croydon has probably been the one that has exploited it most. . . . And I think what it has done is to open up Croydon to this whole corridor. [Interview C12]

Croydon Council and London Transport had for some time been concerned to integrate a large isolated housing development at New Addington with the rest of Croydon. Moreover, there was a recognition of an economic logic to Tramlink, from Croydon Council's perspective, in that it would expand the labour market open to the central Croydon commercial centre, bringing other South London boroughs into the orbit of Croydon in the form of increased commuter flows (Croydon Council, 1992). This latter impetus began to become important in the economic malaise that seemed to have come over Croydon in the early 1990s. Thus,

> One of the reasons for building Tramlink in the very beginning was that the businesses... were complaining that it was becoming increasingly difficult to come into Croydon.... Therefore if something wasn't done, they – particularly the businesses not necessarily the retailers but the commercial side of it – might have to start looking elsewhere. [Interview C13]

Tramlink has indeed expanded the labour markets open to the larger town centre office and retail employers but it has also improved the image of Croydon as a destination for inward investors in these two sectors (Colin Buchanan and Partners, 2003).

Croydon Council was instrumental in establishing the edge cities network from which our case-study post-suburban municipalities are drawn. Croydon Council's championing of the European network and use of the term 'edge city' sit somewhat uncomfortably with longer and more firmly held local beliefs in the borough's being a city in its own right and its relations with central government and the GLA. Croydon first bid for city status in the early 1900s and, according to its latest bid (Croydon Council, 1999), is the largest town in western Europe without city status. However, there were concerns that its bid for city status would compromise the borough's long-standing political connections to central government and jeopardise its position vis-a-vis the GLA [Interview C5]. In this respect, then, Croydon's opportunistic self-styling as an edge city bears little resemblance to the North American idea of an edge city as popularised by Garreau. As one senior economic development officer recounted,

> We didn't think at all about the American concept.... Part of the psychology is that there isn't a psychology underneath it. There's this Croydon as a city... This kind of European city kind of concept

that Croydon has. It wants to punch above its weight. It wants to be something it's not. . . . The interesting thing about edge city is not the edge it's the city. [Interview C9]

These sentiments were reiterated in the context of the body charged with place marketing for Croydon, for whom it was suggested, 'It's the last thing I would want to be known as on the edge of something. I want us to be central to the whole of the South East' [Interview C5].

The championing of this trans-European edge cities network has done as much as anything to raise the profile of the borough and enhance its claim to be a city in its own right (Meikle and Atkinson, 1997). Here, then, Croydon Council has lent weight to its own local political manoeuvring when enlarging its space of engagement through this trans-European local authority network. There is just the hint of the sense in which 'local identity and the urban territory, as a stratified deposit of natural and cultural assets, no longer have value for what they are but for what they become in the process of valorisation' (Dematteis, 2000: 63). To the extent that a transnational local authority network, like the edge cities network, meshes with local political coalitions and partnerships it can lend weight to the latter.

Croydon embodies a post-suburban *place* whose identity has been transformed by the entrepreneurial actions of its major institution, Croydon Council, from a dormitory suburb to a suburban office and commercial centre in the 1960s to a city-in-waiting with a wider independent economic and administrative sphere of influence within South London and beyond (Figure 8.5).

8.5 Business at the margins?

The picture of business interest representation in Croydon (depicted in Figure 8.6) is the most complicated and most well developed among our case studies of post-suburban Europe. In part, this reflects the highly centralised nature of business interest representation in the UK that has seen major business representative bodies at their most organised and active in London. Nevertheless the organisation of business interests centred on London-specific issues has evolved recently to involve several major players, London First, the London Chamber of Commerce and Industry (LCCI) and London CBI (LCBI), alongside the long-established London Corporation each of which have developed strong connections to central government and to pan-London authorities such the GLA,

Figure 8.5 Croydon central office complex and tramlink

LDA and Government Office for London (GoL) [Interview C16]. Significantly, the main agendas of these pan-London business interest groups rarely touch on South London let alone Croydon-specific issues.[2]

The complexity of business interest representation in Croydon also, in no small measure, stems from the long history of substantial development in the borough, the long-standing role of business people in its administrative and political affairs and the style of urban politics

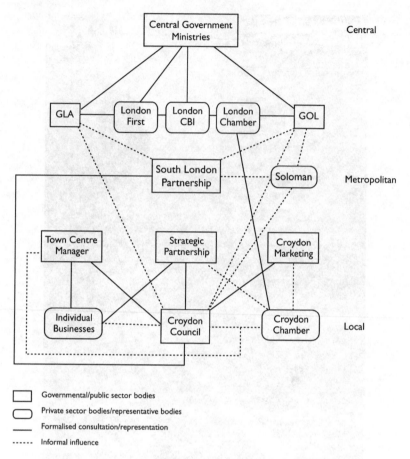

Figure 8.6 Map of business interest representation in Croydon

promoted locally by the Council. In short, Croydon is illustrative of the centrality of significant business interests to urban politics in post-suburbia – a case of business being anything but marginal in the affairs of municipalities at the edge of major city-regions. So, as one interviewee described,

What there is in Croydon is a certain amount of self-containment. There's this feeling of self containment which is not necessarily a sort of hostility to being in London it's simply that the town functions as more of an entity than most outer boroughs. I think you also have this active business community partly because of the relatively long development as a centre of employment. [Interview C3]

The size and strength of Croydon as an employment centre increased dramatically since the initial 1960s office and retail developments so that

> We have a very strong business community here... from our perspective that's another big plus for us from a purely commercial point of view. Other head offices breed affluent customers.... In terms of whether we would like to be in a town that has other national brands, then, yes, it all adds to the flavour if you like. [Interview C6]

As these quotations indicate, the sheer size of the central Croydon economy has meant that there is a strong and politically active business community. It has also, as we will see below, permitted and supported the development of the most developed set of formal and informal channels of business representation at the local level among our case-study post-suburban municipalities.

There are some important continuities in the types of business interests that have historically been marginalised within the local arena. For example, away from the large retail and office-based business interests based in central Croydon other business interests appear marginalised. Since Croydon's reincarnation as a retail commercial centre in the 1960s there has been an emphasis and concentration on servicing the needs of service-sector industries to the neglect of the borough's once significant engineering industries (Saunders, 1983) and a long-term process of closure and relocation out of the borough. Periodically, neglect of manufacturing industry has surfaced in local politics as with London as a whole (Phelps, 1998), but rarely has this interest been sustained in a way that the interests of manufacturers have been incorporated seriously into the workings of government at an all-London, South London or local level. More recently then, the Council has become concerned about retaining manufacturing companies in the borough [Interview C19], while Croydon businesses have played a prominent role in the South London Association of Manufacturers (SOLOMAN) established in the late 1990s. However, as a representative of SOLOMAN complained, 'it's not enough. You've got people fighting globalisation and on the other hand there's the tram, you know!' [Interview C11]. Similarly, Saunders was able to observe how 'there is in Croydon a significant economic and political division in the business community between the centre and the periphery, the big multinationals and retailers on the one hand and the small shopkeepers in the outer areas on the other' (Saunders, 1983: 304). Again, whilst the Council periodically has been sensitised to the needs of diverse business communities outside of the central office-retail complex (Phelps, 1998), these interests have rarely shaped local decision-making to any serious extent.

Yet there are also some important changes in the structure and style of representation of dominant business interests locally. Saunders (1983: 211) traces the influence of business people within Croydon's government back to at least the mid-1800s. During the years in which Croydon grew as a commercial centre in its own right there existed a 'relatively dense and cohesive network of business and political activists, interacting regularly and relatively informally in a variety of institutional contexts' (Saunders, 1983: 313) so that senior managers of major town centre companies were 'intimately engaged in the administrative and political life of the town' (Saunders, 1983: 310). By the 1980s and 1990s after years of informal influence described by Saunders, a formal mechanism had been established to canvass business views by the council's economic and strategic development unit (ESDU) which established the Croydon Economic Development Forum. This was a forum that was unsurprisingly dominated by private-sector interests. Following central government recommendations and rules, it was then merged into the council's Croydon Partnership under which there was equal representation of business, voluntary and public sector. 'But business wasn't happy with that structure . . . In terms of policy and strategy . . . business wasn't interested . . . what they are interested in is impacts' [Interview C14]. As another interviewee commented on his concerns over

> the strategic body if you like of Croydon Plc getting tied into, slightly worryingly to me, the Community Partnership. Because it's drifting away from making wealth and dealing with the economy into looking after the community. . . . It seems to me you have to have the economy right and other things should follow. [Interview C7]

Indeed, the influence of Croydon Chamber as a vehicle for business interest representation appears to have lessened in such institutional reorganisation. In the light of the GLA and LDAs desire to coordinate on a sub-regional basis and in the light of the strong impetus provided to East London business and other interests as a result of the Thames Gateway initiative, Croydon Chamber has been swept up into efforts to coordinate the rationalisation of business voices on a South London basis, in which respect there has been success in comparison to other sub-regions of London [Interview C15].

The Croydon Partnership may have appeared to have diluted business representation and hence the influence in council decision-making that existed in the 1990s; however, Figure 8.6 makes clear that there are now additional powerful vehicles through which business interests are represented to the council and to other local public institutions. The Council's own private marketing arm – Croydon Marketing and Development (CMD) – is clearly important and the preferred channel of influence of at least one interviewee.

> The Council is very keen to have different focus groups and talking shops to keep abreast of what the community wants as a whole. But the very fact they finance CMD to a serious amount of money a year just tells me what they think is the best place to be. [Interview C6]

If 'to spend is to choose' as the adage goes, then CMD represents an important Council-linked decision-making forum which is almost exclusively the preserve of business interests. This quote reveals two important aspects of interest representation in urban politics. First, it suggests that the council is more oriented towards business interests. Second, that the main channel for business interests is via the council's own commercial arm, CMD.

A second major channel of influencing local decision-making on both day-to-day matters and indirectly the Croydon Partnership strategic policy-making vehicle is through the Town Centre Manager. He was able to identify a core set of business interests that repeatedly were involved in local policy-making fora.

> I think it's the same in every town. You always seem to end up with the same faces. And the same agendas keep coming round and round. And that is one of the reasons why I have been trying to get smaller businesses involved because they come up with different issues, or they have different problems, and also some times there are some very unique solutions for them. I mean you can always guarantee at the meeting there will be someone from Nestlé, there will be someone from BT, there will be someone from the Home Office. [Interview C12]

Beyond these new fora for business representation, the Council has also had a dialogue with town centre retailers [Interview C8] and has recently established an 'after-care' programme covering the largest town centre offices [Interview C19]. Whilst there is an interest here in encompassing smaller business interests within the central Croydon business complex, taken together these observations illustrate the multiple and mutually reinforcing direct and indirect channels of business interests to influence local decision-making especially that centring on Council expenditures and strategic priorities. In short,

> The good thing about big business in the town is that they are taken seriously by the Council because they are big rate payers and a lot of the staff of those firms live in the Croydon area. So our input with the Council on what goes on in Croydon is reasonably substantial. [Interview C6]

8.6   Conclusion

More than anything, the case of Croydon highlights the evolution of post-suburbia. By the early 1900s, Croydon had evolved from a physically non-contiguous market town to a residential suburb and part of the continuous urban fabric of greater London. It evolved again to function in economic terms rather like a US edge city with the redevelopment of its central area. Since this time the economic mass of Croydon has been enhanced further and its economic, and also to a lesser extent its administrative and political, impact upon neighbouring boroughs felt in rather the same way that what Lang (2003) terms 'edgeless cities' can sprawl within the existing urban fabric.

Business interests have been important within Croydon from its days as a market town but the Council has played a central role in articulating these interests in local politics surrounding the development process since the 1950s. The sheer mass of economic activity in central Croydon let alone the borough as a whole has produced a growth of business representative bodies and channels for local interest representation. Within this increasingly complex local political scene, the Council has embedded itself more firmly as a central coordinator of local public, private sector and voluntary interests and promoter of development within what could be characterised as a US-style urban regime (Stone, 1989). In a changing pan and sub-London institutional scene, the Council has also begun to expand its spaces of engagement (Cox, 1998) – articulating local and non-local interests and institutional machineries.

9
Post-Suburban Futures

> The city – it is the city that will have to heal them, with knife
> blade and automobile, nightstick, gunshot. The local passions
> of love and hate. The loose cables and rogue masonry of the
> telekinetic city.
>
> Martin Amis, Time's Arrow

9.1 Introduction: The city as collective actor?

In Martin Amis's novel the injuries inflicted by one man upon many
are healed as time runs in reverse. It would be a mistake to think
that turning back of time is any solution to the issues facing modern
cities and their post-suburban edges. Nostalgia and a desire to retain
historic city forms or return to some mythical rural idyll have provided
powerful and enduring ideas shaping urban planning. Yet, one lesson
to be drawn from at least some of the chapters in this book is that a
fondness of the past is, of itself, insufficient to produce the 'good city'.
As we saw with the case of the Finnish municipality of Espoo, Ebenezer
Howard's garden city ideals were an integral part of a deconcentrated
and decentred urban form that has been relatively problematic in this
particular metropolitan and national context. There is a sense then in
which the past – or a critical reconsideration of the past – is important
to the future. As Sandercock (2003: 47) notes, 'in order to imagine the
future differently, we need to start with history, with a reconstruction
of the stories we tell ourselves about planning's role in the modern and
post-modern city'.

If Relph's argument that 'the possibilities for maintaining and reviving
man's sense of place do not lie in the preservation of old places' (1976:
145) is true, then indeed 'in 2003 it is too late to correct past mistakes'

(Hayden, 2003: 229). While the economies of London's suburbs have been relatively robust in the face of new sources of international competition and have not suffered badly in terms of environmental and infrastructural decay (GLA, 2002), it is now apparent that many older US suburbs are in need of revitalisation (Hayden, 2003; Orfield, 2002). The lesson to be drawn from the quotation above is that the city conceived as a collective actor must heal itself. 'The metaphor of the city as actor suggests that actors mobilize within the city to produce strategies, institutionalise contexts for interaction, norms, and collective choices, and that these have an impact on the city's future' (Le Galès, 2002: 224). Le Galès highlights five dimensions of the collective actor nature of cities: (1) common interests within the city; (2) internal and external representation; (3) collective decision-making; (4) integration mechanisms; and (5) capacity for innovation. What these five highlight is the interconnections between civic and political mobilisation around questions of place-identity and administrative implementation and, given the cases presented in previous chapters, the real difficulties in allying the two. Thus one of the key stories to be told about the shaping post-suburbia from the extant literature and this book is simply the degree to which planners and the planning system along with a wide range of other actors – businesses, land owners, property developers, building companies, national and local governments and even residents and environmental groups – are all implicated in the systemic production of the increasingly diffuse urban form that is post-suburbia.

Moreover, if 'urban form has become irremediably dispersed', in what sense can we begin to speak of the city as collective actor? 'If clustering is to be found in cities, it is as an amalgam of institutions, of regulations, power, representation and sociability, some of which are specifically urban (such as local governments), while others are translocal institutions with a presence in cities' (Amin and Thrift, 2002: 73). Amin and Thrift go on to interpret cities as a density of 'light institutions'. In the light of our comments in Chapter 2 and several of our empirical cases, we might argue in contrast that there remains a need to analyse the still significant role of weighty institutions such as the central and local state in urban development. It is clear that interventionist states with sophisticated planning systems as different as Finland (see Chapter 7) and Japan (Sorensen, 1999) have managed to produce quite startling instances of urban sprawl. Notwithstanding this, we have in this book tried to convey something of the complexity of these amalgamations of institutions in the formation and development of post-suburban Europe.

9.2 Collective action and the making of post-suburbia

Our research does indeed confirm that 'cities cannot be read as economic machines, as bounded economic space with special properties of place' but as an 'ecology of circumstance' (Amin and Thrift, 2002: 76–77). It is strange that we should return to an ecological metaphor since one was also so closely associated with the modern metropolis depicted by the Chicago School. Yet, without wanting to return to adopt all its intended and unintended connotations, the metaphor of an ecology of circumstance does highlight at least two important facets of urban growth: the competition between and the meshing of local and non-local processes and institutions; and geographical and temporal disparities within and between national contexts.

Dynamic or state created?

There is a case for arguing that debate regarding contemporary urbanisation is overly imbued with the apparently spontaneous processes of US-style post-suburban growth. Although much of the recent literature of the 'Los Angeles School' has tended to highlight the role of government expenditures and policies in facilitating urban sprawl, we might argue that the analysis of the state's role in promoting post-suburban development needs to go further. This is certainly the case in the European setting and even more so in East Asia where urban development has been integral to national economic development models. Cities, and one might presume post-suburban developments, have served as synonyms for the nation within such state development strategies (Bunnell, 2002; Olds, 2001).

In this book we have only been able to make a crude first-cut at stressing and analysing the role of the state in leading or ostensibly creating new urban spaces at the edge of major cities in Europe. It is surely an analytical theme that can be further developed. The high point of modernist planning may have passed but across Europe and elsewhere the size of the state has yet to show signs of diminishing (Le Galès, 2000), its created spaces present an ongoing commitment, and all too often burden, on the part of government and wider society. The conditions under which such experiments can transcend themselves to become places is surely one that is of considerable moment.

Alternate agents

It is clear that perhaps the major starting point for charting differences and similarities among post-suburban forms must be an analysis of the various agents involved in the process of urban development. Historically, there has been a strong tradition of this analysis within the planning literature (Ambrose, 1994; Healey and Barrett, 1990). Arguably, more of these sentiments ought to be incorporated into the urban geography and urban studies literature which may have over-generalised from particular global settings and post-suburban forms.

Some of the differences in the dominant actors and coalitions of actors involved in the development of post-suburbia in different global settings, such as North America, Europe and East Asia, have only begun to be explored in the extant literature. Moreover, given differences in state constitutions and private-sector industry structures that exist, even where the same actors or groups have led in urban development at the edge of major cities, they may have done so in different ways and with different outcomes in terms of the function and form of post-suburbia. In this way the appearance and wider effects of both firmly state planned and free-market private-sector-led post-suburban developments can differ markedly from their intentions – so that it becomes futile to ascribe particular styles of post-suburban development to any particular agent or coalition of agents.

Some time ago, Castells highlighted the city as 'the result of an endless historical struggle over the definition of urban meaning by antagonistic social actors who oppose their interests, values, and projects...' (Castells, 1983: 335). And as Castells observes in *The City and the Grassroots* there are clear limits to interpreting the city solely in terms of the logic of capital. The relationship between structure and agency remains a key problematic in urban studies. Yet a paradox exists which is barely addressed in the literature. On the one hand, then, Hogan can complain that 'the problem is not the actors – these are the best and the brightest. The problem is the system – the free market in land and development for profit, with plans and planners for sale' (Hogan, 2003: 66). The paradox is that despite the different actors with seemingly different and potentially conflicting interests, there is in reality an accommodation among actors in planning for sprawl. This is what Hogan terms 'big picture' planning in which professionalisation has played a major role in driving a degree of isomorphism. On the other hand, then, modernism may be alive and well it is just not in the exclusive preserve of professional town planners but more widely diffused among a cadre of professional architects, town planners, engineers, property developers and so on who are increasingly geographically and occupationally mobile especially across the public–private sector divide.

Fungible functions

Another purpose of this book has been the desire to shift discussion away from a static view of the functioning of post-suburban developments. The proliferation of terms such as 'edge city', 'technoburb' and 'edgeless cities' is in part a reflection of the fact that much effort has been devoted to distinguishing the functional role of different types of post-suburban forms. Yet such terms may also serve to obscure some important points of comparison and continuity in post-suburbia given that these areas continue to evolve over time. Croydon's evolution is one example of where a relatively established and bounded territory has continued to evolve in function from a dormitory suburb to a more balanced dormitory-employment complex. Even in Espoo it is apparent that the nature of this residential suburb has evolved from a dystopic highpoint of extremely fragmented and under-serviced discrete developments to a higher density group of urban centres for which urban amenities and public transport services and infrastructure are beginning to become viable.

Such changes in function are not simply the product of serendipity since it is apparent that considerable political leadership or vision has been implicated. In the 1960s, Croydon was able to take advantage of market forces and state policies favouring employment decentralisation only in the context of decisive local action on the part of the council and its political leaders. The evolution of Getafe into a more balanced community and 'capital of the south' of Madrid has been driven by local political mobilisation.

Friable forms

The term 'edge city' is now firmly invested with connotations of the physical form of US-style post-suburban development. This alone presents major problems for comparative analysis of urban development at the edge of major cities. Undoubtedly, and as this book has served to underline, there are major differences in the appearance or form of post-suburban developments. As the preceding chapters have illustrated, the form of post-suburban developments in Europe is (a) rather different from nation to nation within Europe, (b) different from the archetypal US-style edge city growth, and (c) different again from any form of East Asian post-suburban development we might care to try to generalise about. Moreover, not only do post-suburban forms vary according to broad geographical region in this way, but they clearly evolve to vary over time in any one geographical setting.

How ought we to investigate such variations in post-suburban form? Our concern in this book has been to try to find an analytical path between, on the one hand, the overgeneralisation visible in the recently declared Los Angeles School, and on the other hand the idiographic tendency to perhaps too readily emphasise the uniqueness of cities in different national and continental settings associated with traditions of area-specialisation within geography. To return to the analytical device of Plato's theory of forms broached in Chapter 4, this actually existing form – perhaps embodied in the archetypal US edge-city example such as Tyson's Corner – is simply one of many variations of the idea of post-suburbia. European variations on this theme are themselves diverse, but paradoxically, because of their diversity highlight some of the parallels in post-suburbia that exist at different geographical scales or with differing degrees of intensity or that have existed at different historical moments in different settings. Espoo came closest to the visual appearance and form of US-style urban sprawl. Kifissia was perhaps the least like a US-style edge city. Yet each of our case-study municipalities either are or have been considered important examples of urban sprawl in the respective national contexts.

Post-suburbia reveals itself as something that is part of the increasingly complex geometry of contemporary city-regions. Post-suburban areas are often internally fragmented or heterogeneous in socio-economic complexion and physical form as well as having important political and economic interactions with neighbouring areas laterally and radially. Certainly post-suburban forms are friable in the sense that they are, as has long been recognised, constantly evolving. Post-suburban agents are actively defining and redefining their territorial boundaries and with it the very identity of the settlements they are connected to. Post-suburbia reveals itself as a form that can in some circumstances be quite porous – representing an amalgam of urban and rural land uses – though at markedly different geographical scales in North America, Europe and East Asia.

9.3 Post-suburban prospects

Planning as an unfettered modernist experiment in rational interventions in, ordering and integration of, land, property and infrastructure markets may have represented a brief interlude in a longer-term history of what is possibly, as we argued in Chapter 2, misconstrued as 'unplanned' urban expansion. According to Sieverts (2003) this 'golden age' – in which an uneasy truce between municipal government and capital prevailed – is now well and truly over. As such, as Le Galès highlights, European cities are only 'collective actors in the making' and one might argue actors whose collective powers are anything but assured.

As such, the development of different city forms has been 'framed by wider efforts to manage and govern urban complexity and conflict' within contexts in which there are 'contested representations of what constitute "order", "chaos", "progress" and the "good city"' (Graham and Marvin, 2000: 389). Town planning practice has tended to be slow to respond to these contested notions. As Sieverts outlines,

> town planning and spatial planning have become defensive, they are attempting to defend the old structures with unsuitable means and they are yearning for old images, without realising that the traditions have become so empty that they are incessantly breaking up... the old traditions and rituals of a halfway homogeneous cultural urban society are dissolving into culturally unassimilated and incompatible individual elements, which can no longer be brought together through an overarching "master plan" in the spirit of traditional urban development. (Sieverts, 2003: 150)

As Jane Jacobs noted some time ago, the growth of cities produces significant challenges to the planning and management of urban space.

> Cities will not be smaller, simpler or more specialized than cities of today. Rather, they will be more intricate, comprehensive, diversified, and larger than today's... The bureaucratised, simplified cities, so dear to present day city planners and urban designers... run counter to the process of city growth and economic development. (Jacobs, 1970: 250–251)

Jacobs's predictions have posed precisely the sorts of challenges to planning implied above. And so, as Healey has recently summarised,

> Planners and plans have been criticized not merely for trying to "order" the dynamic and inherently disorderly development of cities and regions. The concepts that have been used... are seen to reflect a view of geography which assumes... contiguous space... that physical proximity is a primary social ordering principle and that place qualities exist objectively, to be... made by physical development and management projects. (2004: 47)

However, the fact that 'our urban systems are simply too large and complicated for us to understand, let alone fix' (Bruegmann, 2005: 457) does not mean that there is no role for intervention but rather that, seen in perspective, such interventions will be temporary and make-shift in nature. Part of the problem has been that, in the past at least, the interests of virtually all major actors in the suburban and post-suburban development process have coincided. As Gottdiener (1977: 149) notes, 'the partners in the planning process, with the exception of the professional planners, take a very limited view of regional needs'. This would include residents who are implicated in land and property market speculation and typically display a 'limited liability'. As such, 'to shape the landscape, the professionals have to win the support of the people with primary decision-making responsibilities: builders, public officials and home buyers' (Rome, 2001: 268).

Leadership

Part of the problem here appears to be that planners, perhaps like some of the academic discussion of urbanisation, have been overly concerned with the appearances of city form. Thus Sieverts has argued that

> The contemporary discussion is still too much restricted to the form of the city: here the compact European city, there the dissolved American urban sprawl . . . By contrast, a comparison of the political goals and processes of town planning could be productive. (Sieverts, 2003: 154)

Such a focus on the goals and processes of town planning could usefully be augmented by a much wider development of knowledge surrounding the specific histories and culture, problems and prospects of suburban and post-suburban developments. So, for example, 'the rehabilitation of older suburbs needs to build upon social, economic and cultural histories and understandings of vernacular architecture' (Hayden, 2003: 235). Jacobs barely hinted at but Sjoberg made explicit the thought that economic growth in cities was dependent upon the development and agency of political power wielded in cities. Moreover, although looking for regularities in the form and function of the pre-industrial city in a manner that would be objected to in the Los Angeles School, one of the key observations that Sjoberg made was the existence of 'contradictory functional requirements' by which he meant that 'various structural arrangements may be at odds with one another' (Sjoberg, 1960:13). The key point here is that political power and leadership – informed by a fuller knowledge of the life of suburbia and post-suburbia – are important in reconciling these contradictions. 'The shifts in institutional cultures that are required... call for strong, visionary leadership that encourages and rewards exposure to new ideas and risk taking' (Sandercock, 2003: 216)

If the planning of contemporary and future urban forms is no longer something to be achieved through primarily administrative or technocratic means, there is most definitely a continued sense in which it remains the preserve of politics as part of cities as collective actors. As Carver pointed out some time ago now, 'the fine art of politics must take its place in building the new kind of suburbanized city... The whole static structure of local government must be given a new flexibility... to deal with this... growth of cities' (Carver, 1962: 118–119). Politics and leadership are important because as Carver reminds us,

> The settlers on the modern frontier are not like the early pioneers who cleared the chosen land themselves, selected the site for the homestead, and made their own environment. Making the suburbs has been a complex, impersonal, greedy industrial process for converting raw land into a finished salable product. (1962: 118–119)

We might add that the state is deeply implicated in some of the impersonality of this process in certain settings. A key question that continues to dog the planning system is how to harness popular support to broader strategic planning to improve the quality of our city-region scale urban environments. Hence the recent communicative (Forester, 1993), collaborative (Healey, 1997) or deliberative (Forester, 1999) turn in planning. In such approaches the emphasis is upon the decision-making process rather than its outcomes *per se* and tends to downplay the extent to which some form of leadership is required to provide dynamism to the process (Phelps and Twedwr-Jones, 2000). The strong autocratic and undemocratic political leadership of city boss politics can lead to potentially wasteful patterns of urban development seen in the US and some UK cities. Strong autocratic leadership, where linked to a popular mandate, can, as we saw in the case of Getafe, be a force for good. Again there is little point in associating one form of leadership with a particular outcome. The strong, autocratic, leadership of Mayor Pedro Castro in Getafe has been vital to the improvement of Getafe into a more balanced community – a development that can from a local perspective be read quite positively.

9.4   Closer to the edge: Future directions for research

In this book we have attempted to address the subject of patterns and processes of post-suburban development in Europe with wider reference to discussion of developments in North America and East Asia. 'An important element of comparative research is... to present... cases as a set of interrelated economic, political and social processes embedded in an institutional system' (Pierre, 2005: 456). Although there are undoubted omissions in our empirical coverage, we hope that the stories of post-suburban development presented here do indeed capture many of these interrelated processes. Moreover, in situating our cases within their respective national settings, we hope, to a degree, to have illustrated some of the ways in which the rules and resources implicated in national systems of governance provide the antecedents to local agency (Sellers, 2005: 430).

Implicit in our wider referencing within Chapter 2 as a thematic guide to these cases of post-suburban development has been a concern for a more genuinely comparative treatment of those points of comparison and difference that exist in post-suburban developments in different national and continental settings. Comparative treatments are important in theoretical terms – in order to broaden debate regarding the function and form of post-suburbia and the agents involved in its development. Comparative treatments are also important in acting as a corrective to the dominance of US perspectives and a tendency to generalise from this particular vantage point apparent in the fields of urban studies, urban sociology and urban geography.

Some time ago Castells argued that 'we need a theory able to explain how city forms... are produced. At the same time, we need a theoretical perspective flexible enough to account for the production and performance of urban functions and forms in a variety of contexts' (Castealls, 1983: 336). Clearly this book does not aim to advance such a theory. It does, however, argue the need to search for understanding between the general and the specific, between nomothetic and idiographic approaches. To this end it does advance a number of themes that we hope are useful in broadening the debate and in furnishing new points of departure for future research.

Comparative treatments are no less important as we have seen in addressing questions of progressive policies to shape and intervene in the post-suburban sphere. A more critical debate regarding the potential outcomes of liberalised versus interventionist planning systems is needed. Fuller consideration of how different forms of political leadership and a renewal of the capacities of, and powers wielded by, planning systems might influence the quality of the post-suburban environment is also needed.

Finally, in academic terms there is much to be gained from the integration of insights from economics, urban studies, planning and geography. Too often, important insights have not been integrated due to their seeming contradictions. Yet urban sprawl and the post-suburban landscape provide prime examples of the paradoxes of contemporary society – something that is of intrinsic interest and might be a purposeful subject for analysis within interdisciplinary approaches. Such interdisciplinarity is also vital if the sorts of unintended consequences that have littered the history of suburban and post-suburban development are to be recognised and the possibilities for, and limits of, planned interventions are to be understood.

Appendix 1
Interview Sources

North Down

ND1   Officer, North Down Borough Council, North Down, Northern Ireland, 30 April 2002.

Kifissia

K1   Senior Investigator, Quality of Life Department, The Greek Ombudsman, 11 July 2003.
K2   Planning officer, Ministry of Environment, Physical Planning and Public Works, 11 July 2003.
K3   Chairman, Society for the Protection of Kifissia, 10 July 2003.
K4   Chairman of the Commercial Association of Alonia, Kifissia, 8 January 2004.
K5   Chairman, Planning Committee, Municipality of Kifissia, 13 October 2003.
K6   European Affairs Officer, Municipality of Kifissia, 10 July 2003.
K7   Employee Relations Assistant, Tria Epsilon Coca-Cola HBC S.A., Kifissia, 9 July 2003.
K8   European Affairs Officers, Municipality of Kifissia, 4 July 2002.
K9   Human Resources Manager, Tria Epsilon Coca Cola HBC S.A., Kifissia plant, 9 July 2003.
K10   Transportation Manager, Tria Epsilon Coca Cola HBC S.A., Kifissia, 9 July 2003.
K11   Officers, Human Resources Department, Alstom, Kifissia, 9 July 2003.
K12   Vice-mayor of Athens, Athens City Council, 8 July 2002.
K13   Consultant and former chairman of Athens Chamber of Commerce and former Mayor of Kalithea municipality, 3 July 2002.

Getafe

G1   Commercial Director, Getafe Inciativas (GISA), 4 November 2003.
G2   Foreign Investment Analyst, IMADE, Community of Madrid, 10 March 2004.
G3   Director, Fundación de Innovación (Getafe), 7 November 2003.
G4   Planning Officer, Regional Planning Department, Community of Madrid, 9 March 2004.
G5   Felipe García Labrado, Getafe Inciativas (GISA), Director, 4 November 2003.

G6   Legal representative, ACEM-representing the metal sector, legal represent-
     ative, 4 November 2003.
G7   Planning Officer, Planning Department, Municipality of Getafe,
     6 November 2003.
G8   European Affairs Officers, Getafe Inciativas (GISA), 2 March 2000.
G9   Getafe representatives for Chamber of Commerce (Madrid), 5 November
     2003.
G10 Vice-president of Young Business Persons Organisation, 5 November 2003.
G11 Getafe commerial sector representative, 5 November 2003.
G12 Company Director, PLADUX, 7 November 2003.

Noisy-le-Grand

N1   Director, Direction du développement économique et de l'emploi (DDEE),
     Noisy-le-Grand, 19 April 2000.
N2   Manager, Sectors 1 and 2, Epamarne, June 2000.
N3   Officer, Direction départementale de l'équipement, Seine-Saint-Denis, June
     2000.
N4   Officer, Syndicat de l'agglomération nouvelle du Val Maubuée, June 2000.
N5   Representative, Groupama and Chair of the Environment & Security
     Committee of Club Ville-Entreprises, July 2003.
N6   General Secretary, RDI Bank Group, DIAC, July 2003.
N7   Director of Marketing and Communications, Océ-France, July 2003.
N8   Director of the Economic Observatory, Direction du développement
     économique et de l'emploi (DDEE), Noisy-le-Grand, July 2003.
N9   Officer, Direction départementale de l'équipement (DDE), Seine-Saint-
     Denis, June 2003.
N10 Director of Information, Communication and Planning, Epamarne, July
     2003.
N11 President of the Chamber of Commerce and Industry, Seine-Saint-Denis,
     July 2003.
N12 Director, Inward Investment, Comité d'Expansion de la Seine-Saint-Denis,
     July 2003.
N13 Officer, Service économique, Seine-Saint-Denis, July 2003.
N14 Economist, Chamber of Commerce and Industry, Seine-Saint-Denis, April
     2000.

Espoo

E1   Research Officer, Wee Gee Project, Espoo City, 30 October 2003.
E2   Architect, City Planning Department, Helsinki Council, 5 September 2003.
E3   Head of Development Planning, YTV, 29 October 2003.
E4   Retired former Deputy Head of Planning Department, Espoo City Council,
     6 April 2004.
E5   Managing Director, Asuntosäätiö, Espoo, 29 October 2003.
E6   Head of Planning, Planning Department, Espoo Council, 8 September 2003.
E7   Planning Assistant, Planning Department Espoo Council, 8 September
     2003.

E8   Assistant Manager, Espoo Chamber of Commerce, Espoo, 14 November 2002.
E9   Researcher Officers, VTT, 8 September 2003.
E10 Transportation planner, Planning Department, Espoo Council, 8 September 2003.
E11 Director of Economic Development, Espoo City, 3 November 2003.
E12 Corporate Communications representatives, Nokia, Espoo, 29 October 2003.
E13 Director of Culture, Espoo Council, 8 September 2003.
E14 Director, Espoon Yrittajat, 3 November 2003.

Croydon

C1   Planning Officer, Planning, Greater London Authority, 24 February 2004.
C2   Research Officer, SOLOTEC, 23 February 2000.
C3   Senior Planning Officers, Government Office for London, 15 December 2003.
C4   Planning Officer, Planning department, Croydon Council, 4 February 2000.
C5   Head of Marketing and Development, Croydon Marketing and Development, 15 August 2001.
C6   Managing Director, Allders UK Ltd, 24 October 2003.
C7   Chief Executive, Croydon Chamber of Commerce, 14 August 2001.
C8   Town Centre Manager, Economic and Strategic Development Unit, Croydon Council, 23 February 2000.
C9   Head of Economic and Strategic Development Unit (ESDU), Croydon Council, 23 February 2000.
C10 Head of Innovation and Enterprise, Economic and Strategic Development Unit, Croydon Council, 15 August 2001.
C11 Co-director, SOLOMAN (South London Association of Manufacturers), 17 September 2003.
C12 Town Centre Manager, Croydon, 19 August 2003.
C13 Freelance consultant and journalist, 19 September 2003.
C14 Croydon Business Development Partnership Manager, ESDU, Croydon Council, 19 August 2003.
C15 Head of Business and Europe Division and Ian Williams Business Connect Branch, Government Office for London, 27 October 2003.
C16 Director of Communications, London First, 2 February 2004.
C17 Advisor on European Economic Affairs, Croydon Council, 19 April 2000.
C18 Edge City Network representative, Croydon Council, 19 August 2003.
C19 Principal Landscape Architect and Business Liaison and Investment Manager, Croydon Council, 21 August 2000.

Notes

1 Introduction

1. According to Garreau (1991) an edge city must (1) have at least two million square feet of office space, (2) have at least 600,000 square feet of retail space, (3) have more jobs than bedrooms, (4) be perceived as a single place, and (5) have been nothing like a city as recently as 30 years ago.
2. In a 'firstspace' perspective, the urban is viewed as a perceived space of materialised spatial practices. In a 'secondspace' view, the urban is viewed as the conceived space of mental or ideational fields. 'Viewed within these two modes of spatial thinking and epistemology, the spatial specificity of urbanism tends to be reduced to fixed forms' (Soja, 2000: 11).
3. Amin and Thrift (2002) identify six conceptions of community: 'planned communities', 'post-social communities' such as those mediated electronically, 'new forms of sociability' – for example the 'light' social relations obtainable in shopping malls, 'diasporic communities', the 'banal communities' of everyday life and 'communities of sympathy'.
4. We are unable to provide a close historical-cultural (Hayden, 2003) or morphological (Whitehand and Carr, 2001) reading of post-suburbia. Our reading of the role of different agents involved in the construction of post-suburban Europe is also partial with markedly less coverage of, for example, building and finance companies involved.

3 In Search of a European Post-Suburban Identity

1. (1) Developing specific local potential, particularly for the creation of permanent jobs. (2) Improving access to the European market for SMEs... particularly through appropriate techniques for co-operation between firms. (3) Improving the supply of services to SMEs which encourage them to innovate. (4) Establishing and developing resource centres to enhance the value of work and improve the integration of women into economic life. (5) Preserving and improving the environment with a view to sustainable development.
2. Prior to this a subset of the edge city partners had also gained funding for a project relating to environmental sustainability under the Commission's REACTE programme.
3. Croydon's share of RECITE II monies amounted to €289,081 (£178,760 at 11 January 2002 exchange rates) over the 4-year period (Croydon Council, 2001a). For the sake of crude comparison, this figure represented less than 10 per cent of the Single Regeneration Budget funds received from central government in the financial year 1999–2000 (Croydon Council, 2001b).
4. The complementarity between the two elements of the project funded under RECITE was not all that it might have been. Support for SMEs has focused on

high technology firms which is likely to have been at odds with the concern to tackle social exclusion.

5. A major part of the RECITE II funding has been directed towards the establishment of a web-based business directory to facilitate the partnering of SMEs in the participating edge municipalities via subsequent video conferencing and face-to-face meetings (see www.edgecities.eu.org). Whilst most partner municipalities reached their target number of firms signing up to the website, the prevalent view is that deeper and permanent relations between SMEs needed council staff to act as brokers [Interview C10]. Nevertheless, by 2002, 755 companies had registered on the website with a further 4000 businesses either actively involved in transnational business co-operation or having capacity built to do so (Gyro Consulting, 2002: 5).

4   Kifissia: Playground of the Athenians?

1. Law 1337/1983.
2. Laws 1892/1990 and 1947/1991.
3. Law 1515/1985 (Gerardi, 1997: p. 243).
4. This is a reference to Homer's *Odyssey*. Aeolus, warden of the four winds, north, south, east and west bound them into a leather bag which he gave to Odysseus with a warning that the bag should not be opened. Aeolus set Odysseus and his men on their voyage with a favourable west wind but while Odysseus slept his companions, curious of Aeolus's gift, opened the bag. All were blown back to the shores of Aeolia from where they were summarily banished.
5. From 1998 onwards there has been a simplification process and in 2003 new legislation enables these voters to vote at their place of residence but for a politician or mayor in the place where they are registered (Greek Ministry of Internal Affairs, 2003).

5   Getafe: Capital of the Gran Sur

1. Curiously, Kelvinator is one of the brands most closely associated with the 'suburban industrial complex' of post-war United States (Rome, 2001).

6   Noisy-le-Grand: Grand State Vision or Noise about Nowhere?

1. The new town is made up of Mont d'Est, Pavé Neuf, Champy, part of les Richardets, and part of Marnois. The old town is composed of the Centre, Varenne, Coteaux, part of les Richardets and part of Marnois.
2. No *contrat de ville* was signed by Noisy-le-Grand at this time, however.
3. Marne-la-Vallée is spread over 26 *communes* in the *départements* of Seine-Saint-Denis, Seine-et-Marne and Val de Marne. Of the 26 *communes*, only Noisy-le-Grand is in Seine-Saint-Denis.

4. The municipality's actions do appear to have had an effect upon local perceptions. In a survey conducted on behalf of the municipality in 2003, 78 per cent of inhabitants questioned said they were satisfied with living in Noisy-le-Grand, while 69 per cent thought that the town had the necessary qualities to affirm its position as the 'capital of the eastern Parisian area' (*Noisy magazine*, 2003).

7 Espoo: California Dreaming?

1. Helsinki city owns 66 per cent of the land within its territory (Helsinki Planning Department, 2000: 22), while Espoo owns just 31 per cent (City of Espoo, 2003).
2. Prior to the creation of YTV, and its subsequent rationalisation and coordination of public transport in the metropolitan area, there were around 20 private bus companies operating in Espoo [Interview 3].
3. One interviewee, a town planner, commented that regional political sensitivities and rivalries had seen use of the term 'Eastern Espoo' to describe areas such as Tapiola, Ottaniemi and Westend, banished from official language because of what it connoted about proximity to Helsinki [Interview E7].
4. http://English.espoo.fi/xsl.etusivu.asp?path = 5731, accessed on 20 March 2003.
5. We are grateful to Timo Heikkinen for this observation.

8 The Croydonisation of South London?

1. Territorial arms of formal organisations such as the Learning and Skills Council, the Small Business Service, the South London Chamber of Commerce as well as informal partnerships such as the London Wandle Valley Partnership, the South London Economic Development Alliance (SLEDA) are all headquartered in Croydon.
2. The main issues of interest highlighted by one interviewee have been the creation of a mayor and Assembly for London, the Cross Rail project, Congestion charging, the 5th terminal at Heathrow and the Thames Gateway redevelopment – none of which directly involve the South London sub-region in any major way [Interview 16]. See Thornley *et al.* (2005) for greater detail of the efficacy of pan-London business interests in shaping the policy agendas of the GLA.

References

Aesopos, Y. and Simeoforidis, Y. (2001a) 'Modernization processes of the 60s and the 90s', pp. 21–31 in Aesopos, Y. and Simeoforidid, Y. (eds) *Metapolis 2001: The Contemporary Greek City*. Metapolis Press, Athens.

Aesopos, Y. and Simeoforidis, Y. (2001b) 'The contemporary Greek city', pp. 32–63 in Aesopos, Y. and Simeoforididis, Y. (eds) *Metapolis 2001: The Contemporary Greek City*. Metapolis Press, Athens.

Allen, J., Massey, D. and Cochrane (1998) *Rethinking the Region*. Routledge, London.

Alonso, W. (1973) 'Urban zero population growth', *Daedalus* 102: 191–206.

Althubaity, A. and Jonas, A. (1998) 'Suburban entrepreneurialism: Redevelopment regimes and co-ordinating metropolitan development in Southern California', pp. 149–172 in Hall, T. and Hubbard, P. (eds) *The Entrepreneurial City: Geographies of Politics, Regime and Representation*. John Wiley, Chichester.

Ambrose, P. (1994) *Urban Process and Power*. Routledge, London.

Amin, A. and Thrift, N. (2002) *Cities: Reimagining the Urban*. Polity Press, Cambridge.

Amis, M. (1991) *Time's Arrow*. Jonathan Cape, London.

Amourgis, S. (2001) 'Greek cities – is there or not historic continuity', pp. 77–81 in Aesopos, Y. and Simeoforidis, Y. (eds) *Metapolis 2001: The Contemporary Greek City*. Metapolis Press, Athens.

Anteroinen, S. (2003) 'Settlement on the E-frontier', *Nordicum*, issue 5, November, pp. 48–51.

Arias, F. (1991) 'La ciudad del sur', pp. 426–430 in Echenagusia, J. (ed.) *Madrid: Punto Seguido*. Cidor Alfoz, Madrid.

Augé, M. (1995) *Non-Places: Introduction to an Anthropology of Supermodernity*. Verso, London.

Ayuntamiento de Getafe (2002) *Un Libro Blanco Para el Desarollo de la Ciudad*. Ayuntamiento de Getafe, Getafe.

Balaquer, I., Cousin, B., Lemp, I., Cousin, B., Lempière, V. and Toupin, F. (1996) *Noisy-le-Grand: Liaison des quartiers. Recherche d'objectifs et démarche de projet global urbain. Tome 1: Diagnostic*. Projet de fin d'études, Mastère AMUR. Paris/Noisy-le-Grand: Ecole nationale des ponts et chaussées.

Barroz, O. and John, P. (2004) 'The transformation of urban political leadership in western Europe', *International Journal of Urban and Regional Research* 28: 107–120.

Bell, M. and Hietala, M. (2002) *Helsinki: The Innovative City*. Finnish Literature Society and City of Helsinki Urban Facts, Helsinki.

Beauregard, R.A. (1995) 'Edge cities: Peripheralizing the center', *Urban Geography* 16: 708–721.

Benedetti, M. (1986) *Preguntas al Azar*. Nueva Imagen, Buenos Aires.

Bennett, R.J. (1997) 'Administrative systems and economic spaces', *Regional Studies* 31: 323–336.

Berry and McGreal (1995) 'European cities: The interaction of planning systems, property markets and real estate investment', pp. 1–16 in Berry, J. and McGreal, S. (eds) *European Cities, Planning Systems and Property Markets*. Spon Press, London.

Benedetti, M. (1986) *Preguntas al Azar*. Nueva Imagen, Buenos Aires,

Bontje, M. and Burdach, J. (2005) 'Edge cities, European-style: Examples from Paris and the Randstad', *Cities* 22: 317–330.

Brenner, N. (1998) 'Between fixity and motion: Accumulation, territorial organisation and the historical geography of spatial scales', *Environment and Planning D, Society and Space* 16: 459–81.

Brenner, N. (1999) 'Globalisation as reterritorialisation: The re-scaling of urban governance in the European Union', *Urban Studies* 36: 431–451.

Brenner, N. (2002) 'Decoding the newest "metropolitan regionalism" in the USA: A critical overview', *Cities* 19: 3–21.

Brenner, N. (2003) 'Metropolitan institutional reform and the rescaling of state space in contemporary Western Europe', *European Urban and Regional Studies* 10: 297–324.

Bruegmann, R. (2005) 'The flaws of anti-sprawl arguments', pp. 448–459 in Chudaloff, H.P. and Baldwin, P.C. (eds) *Major Problems in American Urban and Suburban History (2nd edition)*. Houghton Mifflin, Boston.

Bunnell, T. (2002) 'Cities for nations? Examining the city-nation-state relation in information age Malaysia', *International Journal of Urban and Regional Research* 26: 284–298.

Calder, S. (2000) 'This is Croydon, gateway to the planet', *The Independent*, 16 September, p. 12.

Carver, H. (1962) *Cities in the Suburbs*. University of Toronto Press, Toronto.

Castealls, M. (1983) *The City and The Grass Roots*. Edward Arnold, London.

Castells, M. and Himanen, P. (2002) *The Information Society and the Welfare State*. Oxford University Press, Oxford.

Castro, P. (1999) 'Getafe siglo XXI: Repensar la ciudad', *Getafe '99 Anuario*. Anuarios Locales de Madrid, Getafe.

CEC (1996) 'Recite II Internal inter-regional co-operation. Article 10 ERDF call for proposals No. 96/13', *Official Journal of the European Communities No. C326/96*, 31st October.

Charlesworth, J. and Cochrane, A. (1994) 'Tales of the suburbs: The local politics of growth in the South East of England', *Urban Studies* 31: 1723–1738.

Cheshire, P. (1999) 'Cities in competition: Articulating the gains from integration', *Urban Studies* 36: 843–864.

Chorianopoulos, I. (2003) 'North-south local authority and governance differences in EU networks', *European Planning Studies* 11: 671–695.

Christophilopoulos, D. (1997) The legislative and organisational framework of urban planning in Greece, pp. 95–115 in Aravantinos, A. I. (ed.) *Urban Planning for a Sustainable Development of the Urban Space* Symetria, Athens (in Greek).

Church, A. and Reid, P. (1996) 'Urban power, international networks and competition: The example of cross-border co-operation', *Urban Studies* 33: 1297–1318.

dal Cin, A., de Mesones, J. and Figueroa, J. (1994) 'Madrid', *Cities* 11: 283–291.

City of Espoo (2003) *City of Espoo Pocket Statistics*. City of Espoo, Espoo.

Club Ville-Entreprises de Noisy-le-Grand (CVE) (1999) *La lettre du Club Ville-Entreprises*, No. 3, January, 1999.

Cole, A. and John, P. (2001) *Local Governance in England and France*. Routledge, London.

Colin Buchanan and Partners (2003) *Economic and Regeneration Impact of Croydon Tramlink*. Final Report to the South London Partnership.

Cox, K. (1998) 'Scales of dependence, spaces of engagement and the politics of scale, or: looking for local politics', *Political Geography* 17: 1–23.

Cox, K. and Mair, A. (1988) 'Locality and community in the politics of local economic development', *Annals of the Association of American Geographers*, 78: 307–325.

Cox, K. and Mair, A. (1991) 'From localized social structures to localities as agents', *Environment and Planning A* 23, pp. 197–213.

Council of Europe (1993) *Major Cities and Their Peripheries: Co-operation and Co-ordinated Management*. Council of Europe, Strasbourg.

Croydon Council (1992) *The Croydon Economy, 1992: A Position Statement*. London Borough of Croydon, Croydon.

Croydon Council (1998) *Croydon 2020: The Thinking and the Vision. Executive Summary*. EDAW, London.

Croydon Council (1999), *Why Croydon? Croydon's Bid for City Status*. London Borough of Croydon, Croydon.

Croydon Council (2001a) 'Croydon's European Programmes: Progress on the edge cities network', *Report to Cabinet*, 29th October, Croydon Council, Croydon.

Croydon Council (2001b) *Annual Accounts: 1999/2000*. Croydon Council, Croydon.

Croydon Council (2005) Census figures for Croydon reported online at: http//www.croydon.gov.uk/councilanddemocracy/localareainformation/Census/2001Census/CIN2.

Croydon Partnership (1998) *Croydon Competitiveness Audit*. The Local Futures Group, London.

Davey, J. (2004) 'Stanhope has plans for Croydon', *The Times*, 6 October, p. 53.

Davis, M. (1996) 'How Eden lost it's garden: A political history of the Los Angeles landscape', pp. 76–105 in Scott, A.J. and Soja, E.W. (eds) *The City: Los Angeles and Urban Theory at the End of the Twentieth Century*. University of California Press, London.

Dear, M. (1996) 'In the city, time becomes visible: Intentionality and urbanism in Los Angeles, 1781–1991', pp. 76–105 in Scott, A. and Soja, E. (eds) *The City: Los Angeles and Urban Theory at the End of the Twentieth Century*. Universiy of California Press, Berkeley, CA.

Dear, M. (2003) 'The Los Angeles school of urbanism: An intellectual history', *Urban Geography* 24: 493–509.

Dear, M. and Flusty, S. (1998) 'Postmodern *urbanism*', *Annals of the Association of American Geographers* 88: 50–72.

Delgado, I, Nieto, L.L. and Lopez, E. (1998) 'Functions and duties of Funcionarios Directivas Locales (local chief officers)', pp. 238–252 in Klausen, K. and Magnier, A. (eds) *The Anonymous Leader: Appointed CEOs in Western Local Government*. Odense University Press, Odense.

Delladetsima, P. and Leontidou, L. (1995) 'Athens', pp. 258–287 in Berry, J. and McGreal, S. (eds) *European Cities, Planning Systems and Property Markets*. Spon Press, London.

Dematteis, G. (2000) 'Spatial images of European urbanisation', pp. 48–73 in Bagnasco, A. and Le Galès (eds) *Cities in Contemporary Europe*. C.U.P., Cambridge.

Dick, H. and Rimmer, P. (1998) 'Beyond the third world city: The new urban geography of southeast Asia', *Urban Studies* 35: 2303–2322.

Dieudonné, P. (1992) *Marne-la-Vallée: Le temps des héritiers*. Editions Autrement, Paris.

Direction du développement économique et de l'emploi (DDEE) (1998) *Noisy éco*, January–March, 1998.

Direction du développement économique et de l'emploi (DDEE) (2000) *Tableau de Bord Economique Année 1999*, DDEE, Noisy-le-Grand.

Direction du développement économique et de l'emploi (DDEE) (2003a)*Tableau de Bord Economique Année 2002*, DDEE, Noisy-le-Grand.

Direction du développement économique et de l'emploi (DDEE) (2003b) *Noisy éco*, January–March, 2003.

Direction du développement économique et de l'emploi (DDEE) (2003c) *Untitled Document*. Ville de Noisy-le-Grand, Noisy-le-Grand.

Dowding, K., Dunleavy, P., King, D., Margetts, H. and Rydin, Y. (1999) 'Regime politics in London local government', *Urban Affairs Review* 34: 515–545.

Edge Cities Network (1996) *Meeting of the Partners*. London Borough of Croydon, Croydon.

Elkin, S. (1987) *City and Regime in the American Republic*. University of Chicago Press, Chicago.

Epamarne/Epafrance (1998) *Présentation*. Epamarne/Epafrance, Noisiel.

Epamarne/Epafrance (1999) *Marne la Vallée*. Epamarne/Epafrance, Noisiel.

Epamarne/Epafrance (2003) *Marne la Vallée en chiffres*, Epamarne/Epafrance, Noisiel.

Ezquiaga, J.M., Cimadevilla, E. and Peribañez, G. (2000) 'The Madrid region', pp. 54–65 in Simmonds, R. and Hack, G. (eds) *Global City Regions: Their Emerging Forms*. Routledge, London.

Fernandez, G. (2000), 'Getafe desarrollará, en los próximos cuatro años, el 80% del suelo industrial de la CAM', *Crónica de Madrid Sur*, 1–3 March 144: p. 19.

Fincher, R., Jacobs, J.M. and Andersen, K. (2002) 'Rescripting cities with difference', pp. 27–48 in Eade, J. and Mele, J. (eds) *Understanding the City: Contemporary and Future Perspectives*. Blackwell, Oxford.

Fishman, R. (1987) *Bourgeois Utopias: The Rise and Fall of Suburbia*. Basic books, New York.

Florida, R. (2004) *The Rise of the Creative Class: And How It's Transforming Work, Leisure, Community and Everyday Life*. Basic Books, New York.

Forester, J. (1993) *Critical Theory, Public Policy and Planning Practice*. State University of New York Press, Albany, NY.

Forester, J. (1999) *The Deliberative Practitioner: Encouraging Participatory Planning Processes*. MIT Press, Cambridge, MA.

Forsyth, A. (1999) *Constructing Suburbs: Competing Voices in a Debate over Urban Growth*. Gordon and Breach Publishers, Sidney.

Friedmann, J. (2000) 'Intercity networks in a globalizing era', in Scott, A.J. (ed.) *Global City-Regions: Trends, Theory, Policy*. O.U.P., Oxford.

Friedmann, J. and Wolfe, G. (1982) 'World city formation: An agenda for research and action', *International Journal of Urban and Regional Research* 15: 269–283.

Garreau, J. (1991) *Edge City: Life on the New Frontier*. Doubleday, New York.

Geneys, W., Ballart, X. and Valarie, P. (2004) 'From "Great" leaders to building networks: The emergence of a new urban leadership in Southern Europe?', *International Journal of Urban and Regional Research* 28: 183–199.

Gent, J.B. (1988) *Croydon: The Story of a Hundred Years*. Croydon Natural History and Scientific Society, Croydon.

Gerardi, K. (1997) 'Masterplans of metropolitan areas: The case of greater Athens', pp. 237–248 in Aravantinos, A.I. (ed.) *Urban Planning for a Sustainable Development of the Urban Space*. Symetria, Athens (in Greek).

Getafe.net (1999) 'Entrevista con Pedro Castro', http://www.getafe.net/getafedigital?LOCAL/local1.html [accessed 24/02/99].

Getimis, P. (1992) 'Social conflicts and the limits of urban policies in Greece', pp. 239–354 in Dunford, M. and Kafkalas, G. (eds) *Cities and Regions in the New Europe*. Belhaven Press, London.

Getimis, P. and Grigoriadou, D. (2004) 'The Europeanization of urban governance in Greece: A dynamic and contradictory process', *International Planning Studies* 9: 5–25.

Ghent Urban Studies Team (1999) *The Urban Condition: Space, Community and Self in the Contemporary Metropolis*. 010 Publishers, Rotterdam.

GLA (2002) *A City of Villages: Promoting a Sustainable Future for London's Suburb's*. Greater London Authority, London.

GLA (2004) *London Plan*. Greater London Authority, London.

Glassman, J. (1999) 'State power beyond the "territorial trap": The internationalization of the state', *Political Geography* 18: 669–696.

Goldsmith, M. and Larsen, H. (2004) 'Local political leadership: Nordic style', *International Journal of Urban and Regional Research* 28: 121–133.

Gordon, P. and Richardson, H.W. (1999) 'Review essay: Los Angeles, city of Angels? No, city of angles', *Urban Studies* 36: 575–591.

Gottdiener, M. (1977) *Planned Sprawl: Private and Public Interests in Suburbia*. Sage, London.

Gottdiener, M. (2002) 'Urban analysis as merchandising: The "LA School" and the understanding of metropolitan development', pp. 159–180 in Eade, J. and Mele, J. (eds) *Understanding the City: Contemporary and Future Perspectives*. Blackwell, Oxford.

Gottdiener, M. and Kephart, G. (1995) 'The multinucleated metropolitan region: A comparative analysis', pp. 31–54 in Kling, Olin and Poster (eds)

Gottman, J. (1961) *Megalopolis*. MIT Press, Cambridge, MA.

Graham, S. and Marvin, S. (2000) *Splintering Urbanism*. Routledge, London.

Gravier, J.-F. (1947) *Paris et le désert français*, Flammarion, Paris.

Greek Ministry of Internal Affairs (2003) 'Voting for "heterodimotes": A brief guide, on-line document', http://www.ypes.gr/FAQ_ETEROD.htm (accessed 25 October 2005) (in Greek).

Groupe d'Etudes de Démographie Appliquée (GEDA) (2003) *Analyse Sociodémographique de la Ville de Noisy-le-Grand*. GEDA, Paris.

Gyford, J. (1994) 'Politics and planning in London', pp. 71–89 in Simmie, J. (ed.) *Planning London*. UCL Press, London.

Gyro Consulting (2002) *Review of RECITE II Edge Cities Network Project: Final Report*. Gyro Consulting, London.

Hall, P. (2002) *Cities of Tomorrow*. Blackwell, Oxford.

Harding, A. (1991) 'The rise of urban growth coalitions UK style?', *Environment & Planning C, Government & Policy* 9: 295–317.

Harding, A. (1997) 'Urban regimes in a Europe of the cities?', *European Urban and Regional Studies* 4: 291–314.

Harris, O. (1993) *Cranes, Critics and Croydonisation: The Reshaping of Central Croydon, 1935–1970*. unpublished mimeograph.

Harris, R. (1999) 'The making of American suburbs, 1900–1950: A reconstruction', pp. 91–110 in Harris, R. and Larkham, P. (eds) *Changing Suburbs*. Spon Press, London.

Harris, R. and Larkham, P. (1999) 'Suburban foundation, form and function', pp. 1–31 in Harris, R. and Larkham, P. (eds) *Changing Suburbs: Foundation, Form and Function*. Spon Press, London.

Harvey, D. (1989) 'From managerialism to entrepreneurialism: The transformation of urban governance in late capitalism', *Geografiska Annaler* 71B: 3–17.

Hayden, D. (2003) *Building Suburbia: Greenfields and Urban Growth, 1820–2000*. Pantheon, New York.

Healey, P. (1997) *Collaborative Planning: Shaping Places in Fragmented Societies*. MacMillan, London.

Healey, P. (2004) 'The treatment of space and place in the new strategic spatial planning in Europe', *International Journal of Urban and Regional Research* 28: 45–67.

Healey, P. and Barrett, S.M. (1990) 'Structure and agency in land and property development processes – some ideas for research', *Urban Studies* 27: 89–104.

Hebbert, M. (1992) 'Governing the capital', pp. 134–148 in Thornley, A. (ed.) *The Crisis of London*. Routledge, London.

Hebbert, M. (2000) 'Transpennine: Imaginative geographies of an interregional corridor', *Transactions of the Institute of British Geographers* 25: 379–392.

Heitkamp, T. (2000) 'The integration of unplanned towns at the periphery of Madrid: The case of Fuenlabrada', *Habitat International* 24: 213–220.

Hellenic Office for National Statistics (2004), Census of Greek Population, available online from: http://www.statistics.gr

Helsinki Planning Department (2000) *Helsinki Urban Guide*. City of Helsinki, Helsinki.

Herrschell, T. and Newman, P. (2003) *Governance of Europe's City Regions: Planning, Policy and Politics*. Routledge, London.

Hidalgo, S. (2003) 'PSOE e IU de Getafe dán Carpetazo a la investigacion del urbanismo', *El País*, 7th November, p. 7.

Hill, E. and Wolman, H. (1997) 'Accounting for change in income disparities between US central cities and their suburbs', *Urban Studies* 34: 43–60.

Hise, G. (1997) *Magnetic Los Angeles: Planning the Twentieth Century Metropolis*. Johns Hopkins University Press, Baltimore.

Hodson, R. (1971) 'Croydon: S.London's mini Manhattan', *Financial Times*, 4th June, p. 14.

Hogan, R. (2003) *The Failure of Planning: Permitting Sprawl in San Diego Suburbs, 1970–1999*. Ohio State University Press, Columbus.

Holman, O. (1996) *Integrating Southern Europe: Expansion and the Transnationalisation of Spain*. Routledge, London.

Holston, J. (2000) 'Urban citizenship and globalization', in Scott, A.J. (ed.) *Global City-Regions: Trends, Theory and Policy*. O.U.P., Oxford.

Ile-de-France Regional Council (2000a) *Contrat de Plan Etat-Région Ile-de-France 2000/2006*, Conseil Régional Ile-de-France, Paris.

Ile-de-France Regional Council (2000b) *Contrat de Plan Etat-Région Ile-de-France 2000/2006*. Rapport pour le Conseil Régional présenté par M. Jean-Paul Huchon, Président du Conseil Régional d'Ile-de-France, Conseil Régional Ile-de-France, Paris.

Illmonen, M., Hirvonen, J., Korhonen, H., Knuuti, L. and Lankinen, M. (2000) *Rauhaa ja Karnevaaleja: Tieto – ja Taitoammattilaisten asumistavoiteet Helsingin Seudulla.* Helsinki University of Technology, Ottaniemi.

Imrie, R. and Raco, M. (1999) 'How new is the new urban governance? Lessons from the United Kingdom', *Transactions of the Institute of British Geographers* 24: 45–63.

Institut national des statistiques et études économiques (INSEE) (1999) *Recensement général de la population,* INSEE, Paris.

Jacobs, J. (1970) *The Economy of Cities.* Jonathan Cape, London.

Jessop, B. (1990) *State Theory: Putting Capitalist States in their Place.* Blackwell, Oxford.

Jessop, B. (1999) 'Reflections on globalisation and its (il)logics', pp. 19–37 in Dicken, P., Olds, K., Kelly, P. and Yeung, H. *Globalisation and the Asia Pacific: Contested Territories.* Routledge, London.

Jonas, A. (1994) 'The scale politics of spatiality', *Environment & Planning D, Society & Space* 12: 257–264.

Jonas, A.E.G. (1999) 'Making edge city: Post-suburban development and life on the frontier in southern California', pp. 202–221 in Harris, R. and Larkham, P. (eds) *Changing Suburbs.* Spon Press, London.

Jones, M. (1997) 'Spatial selectivity of the state: The regulationist enigma and local struggles over economic governance', *Environment and Planning A* 29: 831–864.

Karavia, M. (1988) *Kifissia: Aspects of its Beauty and its Past.* Association for the Protection of Kifissia, Kifissia.

Keating, M. (1997) 'The invention of regions: Political restructuring and territorial government in Western Europe', *Environment & Planning C, Government & Policy* 15: 383–393.

Keil, R. (1994) 'Global sprawl: Urban form after Fordism', *Environment & Planning D, Society & Space* 12: 31–36.

Keil, R. (2000) 'Governance restructuring in Los Angeles and Toronto: Amalgamation or seccession', *International Journal of Urban and Regional Research* 24: 758–781.

Keil, R. and Ronneberger, K. (1994) 'Going up country: Internationalization and urbanization on Frankfurt's northern fringe', *Environment & Planning D, Society & Space* 12: 137–166.

Klausen, K.K. and Magnier, A. (1998a) (eds) *The Anonymous Leader: Appointed CEOs in Western Local Government.* Odense University Press, Odense.

Klausen, K. and Magnier, A. (1998b) 'The new mandarins of western local government – contours of a new professionl identity', pp. 265–284 in Klausen, K. and Magnier, A. (eds) *The Anonymous Leader: Appointed CEOs in Western Local Government.* Odense University Press, Odense.

Kling, R. Olin, S. and Poster, M. (1995) 'Beyond the edge: The dynamism of postsuburban regions', pp. vii–xx in Kling, R., Olin, S. and Poster, M. (eds) *Postsuburban California: The Transformation of Orange County Since World War II.* University of California Press, Berkeley, CA.

Koppel, B. (1991) 'The urban-rural dichotomy re-examined: Beyond the ersatz debate', pp. 47–70 in Ginsborg, N., Koppel, B. and McGee, T. (eds) *The Extended Metropolis: Settlement Transition in Asia.* University of Hawaii Press, Honolulu.

Kunstler, J.H. (1993) *The Geography of Nowhere: The Rise and Decline of America's Man-Made Landscape*. Simon and Shuster, London.

Laakso, S. and Keinanen, O. (1995) 'Helsinki', pp. 121–137 in Berry, J. and McGreal, S. (eds) *European Cities, Planning Systems and Property Markets*. Spon Press, London.

Lambert, C., Griffiths, R., Oatley, N., Taylor, N. and Smith, I. (n.d.) 'On the edge: The development of Bristol's North fringe', *Working Paper 9*, Cities Research Centre, Faculty of the Built Environment, University of the West of England.

Lang, E. (2003) *Edgeless Cities: Exploring the Elusive Metropolis*. Brookings Institution Press, Washington DC.

Lefevre, C. (1998) 'Metropolitan government and governance in Western countries: A critical review', *International Journal of Urban and Regional Research* 22: 9–25.

Le Galès, P. (1998) 'Regulations and governance in European cities', *International Journal of Urban and Regional Research* 22: 482–506.

Le Galès, P. (2000) 'Private sector interests and urban governance', pp. 178–197 in Bagnasco, A. and Le Galès, P. (eds) *Cities in Contemporary Europe*. C.U.P., Cambridge.

Le Galès, P. (2002) *European Cities: Social Conflicts and Governance*. O.U.P., Oxford.

Leitner, H. and Sheppard, E. (1999), 'Transcending interurban competition: Conceptual issues and policy alternatives in the European Union', pp. 227–243 in Jonas, A.E.G. and Wilson, D. (eds) *The Urban Growth Machine: Critical Perspectives, Two Decades Later*. State University of New York, Albany.

Leitner, H. and Sheppard, E. (2002) ' "The city is dead, long live the net": Harnessing European interurban networks for a neoliberal agenda', *Antipode* 34: 495–518.

Leontidou, L. (1990) *The Mediterranean City in Transition: Social Change and Urban Development*. Cambridge University Press, Cambridge.

Lewis, R. (1999) 'Running rings around the city: North American industrial suburbs, 1850–1950', pp. 146–167 in Harris, R. and Larkham, P. (eds) *Changing Suburbs*. Spon Press, London.

Logan, J. and Molotch, H. (1987) *Urban Fortunes: The Political Economy of Place*. University of California Press, London.

London Borough of Croydon (1969) *Croydon Official Guide*. Home Publishing Co., Carshalton.

Maldonado, J.L. (2002) 'Metropolitan government and development strategies in Madrid', pp. 359–374 in Kreukels, A., Salet, W. and Thornley, A. (eds) *Metropolitan Governance and Spatial Planning*. Taylor & Francis, London.

Manninen, A. (2003a) 'Espoo's idealistic model city turns fifty', *Helsingin Sanomat* (International edition), 5th August (http://www.Helsinki-hs.net/news.asp?id=20030805IE3 accessed 15/8/03).

Manninen, A. (2003b) 'Gentrification hit's a suburb designed for ordinary people', *Helsingen Sanomat* (International edition), 5th August (http://www.Helsinki-hs.net/news.asp?id=20030805IE2 accessed 15/8/03).

Marks, G., Hooghe, L. and Blank, K. (1996) 'European integration in the 1980s: state-centric v. multi-level governance', *Journal of Common Market Studies* 34: 341–378.

Marriott, O. (1967) *The Property Boom*. Hamish Hamilton, London.

Maspero, F. (1994) *Roissy Express: A Journey Through the Paris Suburbs*. Verso, London.

Mayer, M. (1995) 'Post-fordist city politics', in Amin, A. (ed.) *Post-Fordism: A Reader*. Blackwell, Oxford.

Mayer, M. (2000) 'Social movements in European cities: Transitions from the 1970s to the 1990s', pp. 131–152 in Bagnasco, A. and Le Galès, P. (eds) *Cities in Contemporary Europe*. C.U.P., Cambridge.

Mazierska, E. and Rascaroli, L. (2003) *From Moscow to Madrid: Postmodern Cities, European Cinema*. I.B. Tauris, London.

McGee, T. (1991) 'The emergence of *desakota* regions in Asia: Expanding a hypothesis', pp. 3–25 in Ginsborg, N., Koppel, B. and McGee, T. (eds) *The Extended Metropolis: Settlement Transition in Asia*. University of Hawaii Press, Honolulu.

Meikle, J. and Atkinson, D. (1997) 'Self-made Croydon revamps its image as glittering city of Europe', *Guardian*, 25th June, p. 9.

Merchant, P. (1994) *House of Commons Minutes of Evidence Taken Before the Comittee on the Croydon Tramlink Bill*, 24th February, Document 13.

Meyer, J., Boli, J., Thomas, G. and Ramirez, F. (1997) 'World society and the nation state', *American Journal of Sociology* 103: 144–181.

Molotch, H. (1976) 'The city as a growth machine: Towards a political economy of place', *American Journal of Sociology* 82: 309–332.

Molotch H., Freudenburg W. and Paulsen K.E. (2000) 'History repeats itself, but how? City character, urban tradition, and the accomplishment of place', *American Sociological Review* 65: 791–823.

Muller, P. (1981) *Contemporary Suburban America*. Prentice Hall, Englewood Cliffs.

Neuman, M. (1997) 'Images as institution builders: Metropolitan planning in Madrid', pp. 77–94 in Healey, P. *et al.* (eds) *Making Strategic Spatial Plans: Innovation in Europe*. UCL Press, London.

Neuman, M. and Gavinha, J. (2005) 'The planning dialectic of continuity and change: The evolution of metropolitan planning in Madrid', *European Planning Studies* 13: 985–1012.

Newman, P. (2000) 'Changing patterns of regional governance in the EU', *Urban Studies* 37: 895–908.

Newman, P. and Thornley, A. (1997) 'Fragmentation and centralisation in the governance of London: Influencing the urban policy and planning agenda', *Urban Studies* 34: 967–988.

Newman, P. and Thornley, A. (1997) *Urban planning in Europe*. Routledge, London.

Newman, P. and Thornley, A. (2005) *Planning World Cities: Globalization and Urban Politics*. Palgrave, Basingstoke.

Nikula, R. (2003) 'Housing policy, architecture and the everyday', pp. 116–143 in Tuomi, T. (ed.) *Tapiola: Life and Architecture*. Espoo City, Espoo.

Noin, D. and White, P. (1997) *Paris*. Wiley, Chichester.

Noisy-le-Grand and Villiers-sur-Marne (1999) *Contrat de Ville les Portes de Paris*, Noisy-le-Grand and Villiers-sur-Marne.

Noisy magazine (1999) 'L'Etat s'engage pour Noisy', *Noisy magazine No. 44*, November 1999, pp. 11–15.

Noisy magazine (2001a) 'Horizon Paris ou la nouvelle génération de parc d'affaires', *Noisy magazine No. 60*, April 2001, pp. 8–9.

Noisy magazine (2001b) 'Un vrai centre-ville pour Noisy', *Noisy magazine No. 67*, December 2001, pp. 11–15.

Noisy magazine (2003) 'Les noiséens aiment leur ville!', *Noisy magazine No. 84*, June 2003, pp. 8–9.

Nomoi (2002), *The Economic and Social Structure of 52 Prefectures and 13 Regions*, All Media Publications, Athens (in Greek).

Nuissl, H. and Rink, D. (2005) 'The "production" of urban sprawl in eastern Germany as a phenomenon of post-socialist transition', *Cities* 22: 123–134.

OECD (2003) *Helsinki, Finland*. OECD, Paris.

OECD (2004) *Athens, Greece*. Organisation for Economic Cooperation and Development, Geneva.

Olds, K. (2001) *Globalization and Urban Change: Capital, Culture and Pacific Rim Mega-Projects*. Oxford University Press, Oxford.

Orfield, M. (2002) *Metropolitics: The New Suburban Reality*. Brookings Institution, Washington DC.

Paasi, A. (1991) 'Deconstructing regions: Notes on the scales of spatial life', *Environment & Planning A* 23: 239–256.

Paasi, A. (1996) *Territories, Boundaries and Consciousness: The Changing Geographies of the Finnish-Russian Border*. Wiley, Chichester.

Paasi, A. (2000) 'Europe as a social process and discourse: Considerations of place, boundaries and identity', *European Urban and Regional Studies* 8: 7–28.

Papamichos, N. (2001) 'From the "antiparochi" to the stock exchange', pp. 82–85 in Aesopos, Y. and Simeoforididis, Y. (eds) *Metapolis 2001: The Contemporary Greek City*. Metapolis Press, Athens.

Pascoe, D. (2001) *Airspaces*. Reaktion, London.

Patsouratis, V. (1994) 'The economics of Local Government', pp. 381–409 in Tatsos, N. (ed.) *Fiscal Economics in Greece*, Sbilias, Athens (in Greek).

Peck, J. and Tickell, A. (1994), 'Jungle law breaks out: Neoliberalism and global-local disorder', *Area* 26: 4. 317–326.

Perkmann, M. (2003) 'Cross border regions in Europe: Significance and drivers of regional cross-border co-operation', *European Urban and Regional Studies* 10: 153–171.

Perkmann, M. and Sum, N.-L. (eds) (2002) *Globalisation, Regionalisation and Cross-Border Regions*. Palgrave, Houndsmills.

Phelps, N.A. (2004) 'Clusters, dispersion and the spaces in between: For an economic geography of the banal', *Urban Studies* 41: 971–989.

Phelps, N.A. (1998) 'On the edge of something big: Edge city economic development in Croydon', *Town Planning Review* 69: 441–465.

Phelps, N.A. and Tewdwr-Jones, M. (2000) 'Scratching the surface of collaborative and associative governance: Identifying diversity of social action in institutional capacity building', *Environment & Planning A* 32: 111–130.

Phelps, N.A., McNeill, D. and Parsons, N. (2002) 'In search of an European identity: Trans-European local authority networking', *European Urban and Regional Studies* 9: 112–124.

Pierre, J. (2005) 'Comparative urban governance: Uncovering complex' *Urban Affairs Review* 40: 446–462.

Pincetl, S. (2004) 'The preservation of nature at the urban fringe', pp. 225–254 in Wolch, J., Pastor, M. and Dreir, P. (eds) *Up Against the Sprawl: Public Policy and the Making of Southern California*. University of Minnesota Press, Minneapolis.

Rees, N. (1997) 'Inter-regional co-operation in the EU and beyond', *European Planning Studies* 5: 385–406.

Relph, E. (1976) *Place and Placelessness*. Pion, London.

224   *References*

Richardson, A. and Simpson, E. (2003) *Draft London Plan Examination in Public: Panel Report*. Available at: http://www.london.gov.uk/mayor/strategies/ sds/eip_report/panel_report_all.pdf.

Rochefort, C. (1961) *Les petits enfants du siècle*, Grasset, Paris.

Rodriguez, M. (2003) 'Madrid, camino de Los Ángeles', *La Vanguardia*, 13 July, pp. 4–5.

Rome, A. (2001) *The Bulldozer in the Countryside: Suburban Sprawl and the Rise of American Environmentalism*. Cambridge University Pres, Cambridge.

Russell, B. (1946 [1988]) *The History of Western Philosophy*. Unwin, London.

Sánchez González, M. (1989) *De Alarnes a Getafe*. Ayuntamiento de Getafe, Getafe.

Sandercock, L. (2003) *Cosmopolis II: Mongrel Cities in the 21st Century*. Continuum, London.

Santos, J., Ringrose D. and Segura, C. (2000) *Madrid. Historia de una Capital*, Alianza Editorial, Madrid.

Sarkis, H. (2001) 'On the beauty of Athens', pp. 153 – in Aesopos, Y. and Simeoforidid, Y. (eds) *Metapolis 2001: The Contemporary Greek City*. Metapolis Press, Athens.

Sassen, S. (1994) *Cities in a World Economy*. Pine Forge Press, London.

Saunders, P. (1983) *Urban Politics: A Sociological Interpretation*. Hutchinson, London.

Savitch, H.V. (1995) 'Straw men, red herrings...and suburban dependency', *Urban Affairs Review* 31: 175–179.

Scott, J.C. (1998) *Seeing Like a State: How Schemes to Improve the Human Condition Have Failed*. Yale University Press, London.

Sebald, W.G. (1998) *The Rings of Saturn*. The Harvill Press, London.

Sellers, J. (2005) 'Re-placing the nation: An agenda for comparative urban politics', *Urban Affairs Review* 40: 419–445.

Shakespeare, W. (1970) *A Midsummer Night's Dream (The New Penguin Shakespeare)*. Penguin, Harmondsworth.

Sieverts, T. (2003) *Cities Without Cities: An Interpretation of the Zwischenstadt*. Routledge, London.

Simoni-Lioliou, M. (2002), *Archontisa Kifissia: The romantic history of the past, up to 1950*. Ioannis Grigorakos, Kifissia (in Greek).

Sinclair, I. (2002) *London Orbital: A Walk Around the M25*. Granta, London.

Sjoberg, G. (1960) *The Preindustrial City: Past and Present*. The Free Press, Glencoe, Illinois.

Smith, N. (1992) 'Geography, difference and the politics of scale', in Doherty, J., Graham, E. and Malek, M. (eds) *Postmodernism and the Social Sciences*. MacMillan, London.

Soja, E. (2000) *Postmetroplis: Critical Studies of Cities and Regions*. Blackwell, Oxford.

Soja, E. and Scott, A. (1996) 'Introduction to Los Angeles: City and region', pp. 1–21 in Scott, A. and Soja, E. (eds) *The City: Los Angeles and Urban Theory at the End of the Twentieth Century*. University of California Press, Berkeley, CA.

Sokos, P. (1998) 'Legislation: First flavour of municipal elections – 17 changes in our vote', *Eleftherotypia* (newspaper), 28 February 1998 (in Greek) (online: http://archive.enet.gr/1998/02/28/on-line/fpage.htm).

Sorensen, A. (1999) 'Land readjustment, urban planning and urban sprawl in the Tokyo metropolitan area', *Urban Studies* 36: 2333–2360.

Stone, C. (1989) *Regime politics: Governing Atlanta, 1946–1988*. University of Kansas Press, Lawrence.

Sudjic, D. (1993) 'The edge city', pp. 4–6 in *Croydon: The Future*. Supplement to *Blueprint* magazine, published to coincide with the Croydon the Future exhibition. Croydon, 30th September–10th December.

Sundman, M. (1991) 'Urban planning in Finland after 1850', pp. 60–115 in Hall, T. (ed). *Planning and Urban Growth in the Nordic Countries*. E & FN Spon, London.

Swyngedouw, E. (1997) 'Neither global nor local : 'Glocalisation' and the politics of scale' in Cox, K. (ed.) *Spaces of Globalisation – Reasserting the Power of the Local*, pp. 137–66. Guildford Press, New York.

Syrett, S. and Baldock, R. (2003) 'Reshaping London's economic governance: The role of the London Development Agency', *European Urban and regional Studies* 10: 69–86.

Taylor, P.J. (1999) 'Places, spaces and Macy's: Place-space tensions in the political geography of modernities', *Progress in Human Geography* 23: 7–26.

Taylor, P. (2004) *World City Network: A Global Urban Analysis*. Routledge, London.

Teaford, J. (1997) *Post-Suburbia: Government and Politics in the Edge Cities*. Johns Hopkins University Press, Baltimore.

Thornley, A., Rydin, Y., Scanlon, K. and West, K. (2005) 'Business privilege and the strategic planning agenda of the Greater London Authority', *Urban Studies* 42: 1947–1968.

Tornroos, D. (2004) 'Business to follow skilled workforce in Espoo of the future', *Espoo Hitech*, annual publication of the Espoo Chamber of Commerce, pp. 6–7.

Tuomi, T. (1992) *Tapiola: A History and Architectural Guide*. City of Espoo, Espoo.

Tuomi, T. (2003) 'Tapiola – Garden City', pp. 7–28 in Tuomi, T. (ed.) *Tapiola: Life and Architecture*. City of Espoo, Espoo.

Vaattovaara, M. and Kortteinen, M. (2003) 'Beyond polarisation versus professionalisation? A case study of the development of the Helsinki region, Finland', *Urban Studies* 40: 2127–2145.

Van den Berg, L., Braunn, E. and van Winden, W. (2001) *Growth Clusters in European Metropolitan Cities*. Ashgate, Aldershot.

Van der Veen, A. (1993) 'Theory and practice of cross-border co-operation of local governments: The case of EUREGIO between Germany and the Netherlands', in Cappellin, R. and Batey, P. (eds) *Regional Networks, Border Regions, and European Integration*. Pion, London.

Vardas, A (2004) 'Hotel "Semiramis" Kefalari Square', *Kifissia* (newspaper), 28 September 2004, issue 747, p. 3 (in Greek).

Verdú, V. (2003) 'Madrid, a monster in the making', *El País* (English version) 27th October, p. 4.

Ville de Noisy-le-Grand (2003) *Guide de la Ville*, Ville de Noisy-le-Grand, Noisy-le-Grand.

Ville de Noisy-le-Grand (2004) http://www.ville-noisylegrand.com/Menu.asp?MenuID=2&SmenuID..., accessed 4 October 2004.

Walker, R. (1981) 'A theory of suburbanization: Capitalism and the construction of urban space in the United States', pp. 383–429 in Dear, M. and Scott, A. (eds) *Urbanization and Urban Planning in Capitalist Societies*. Methuen, London.

Walker, R. and Lewis, R.D. (2001) 'Beyond the crabgrass frontier: Industry and the spread of North American cities, 1850–1950', *Journal of Historical Geography* 27: 3–19.

Ward, K. (1996) 'Rereading urban regime theory: A sympathetic critique', *Geoforum* 27: 427–438.

Ward, S. (1998) 'Place marketing: A historical comparison of Britain and North America', pp. 31–54 in Hall, T. and Hubbard, P. (eds) *The Entrepreneurial City: Geographies of Politics, Regime and Representation*. John Wiley, Chichester.

Ward, S. (2005) 'A pioneer "global intelligence corp"? The internationalisation of planning practice, 1890–1939', *Town Planning Review* 76: 119–141.

Weinstein, R.S. (1996) 'The first American City', pp. 22–46 in Scott, A.J. and Soja, E. (eds) *The City: Los Angeles and Urban Theory at the End of the Twentieth Century*. University of California Press, London.

Whitehand, J.W.R. and Carr, C.M.H. (2001) *Twentieth Century Suburbs: A Morphological Approach*. Routledge, London.

Wolch, J., Pastor, M. and Dreir, P. (2004) 'Introduction: Making Southern California – Public policy, markets and the dynamics of growth', in Wolch, J., Pastor, M. and Dreir, P. (eds) *Up Against the Sprawl: Public Policy and the Making of Southern California*. University of Minnesota Press, Minneapolis.

Wolfe, T. (1998) *A man in Full*. Jonathan Cape, London.

Wood, A. (2004) 'Domesticating urban theory? US concepts, British cities and the limits of cross-national applications', *Urban Studies* 41: 2103–2118.

Young, K. and Garside, P.L. (1982) *Metropolitan London: Politics and Urban Change 1837–1981*. Edward Arnold, London.

YTV (2002) *Helsinki Metropolitan area Transport System Plan 2002*. YTV, Helsinki.

Index

Aalto, A., 156
Aesopos, Y. and Simeoforidis, Y., 71, 77–9
Allen, J. *et al*, 38
Alonso, W., 30
Althubaity, A. and Jonas, A., 26, 28, 39
Ambrose, P., 23, 40, 200
Amin, A. and Thrift, N., 198, 199, 211(n3)
Amis, M., 197
Amourgis, S., 79
Anteroinen, S., 170
Arias, F., 106
Association for the Protection of Kifissia, 93
Athens, 13–14
Augé, M., 14, 34

Balaquer, I. *et al*, 121, 140
Barroz, O. and John, P., 107, 118
Beauregard, R.A., 17
Bell, M. and Hietala, M., 150
Bennett, R.J., 19
Berry and Mc Greal, 13
Bontje, M. and Burdach, J., 29, 35
boundary regions, 18, 19
Brenner, N., 17, 18, 19, 20, 33
and Smith, 11
Bruegmann, R., 22, 204
Bunnell, T., 198

Calder, S., 185
CAM, 100–3, 105, 106, 107, 108, 109, 111–15
Carver, H., 205
Castells, M., 32, 97, 99, 100, 200, 206
and Himanen, P., 148–9
Castro, P., 57, 107–9, 110, 111–15, 117
Çatal Hüyük, 1–2, 4
Charlesworth, J. and Cochrane, A., 12, 17
Cheshire, P., 49

Chicago school, 37
Chorianpoulos, I., 45, 48
Christophilopoulos, D., 72, 73, 75
Church, A. and Reid, P., 47, 49, 56, 66
citizenship, 31–2
city-region, and central city/edge city link, 28
concentration on central areas, 17
emergence of, 17
and rescaling of state structures/practices, 20–1
and role of post-suburban areas, 17–18
Cole, A. and John, P., 129
community, 11, 24, 211(n3)
Council of European Municipalities and Regions (CEMR), 44
Cox, K., 12, 27, 39, 186, 196
and Mair, A., 27, 139
cross-border regions (CBRs), 18, 43
Croydon, 15, 42, 49, 51, 52, 53, 59, 61, 196, 211(n2)
architectural design in, 178–9
background, 172–3
business at the margins, 190–6
comparison with USA, 183, 189–90
contrariness of, 173–7
and Croydonisation of South London, 183–90
as edge city, 64–5, 175, 185, 189–90
employment in, 175
and fragmentation of identity, 185
geographic fragmentation, 184–5
independence of, 182–3
internal fragmentation of, 177
as outside/independent of London–wide government, 173–4
and pan-London business interests, 190–1, 213(n2)
partnerships, 186–7
as post-suburban place, 177, 185

227